Macromolecular Symposia 219

Talking about Biocolloids

Anaheim, USA
March 28–April 1, 2004

Symposium Editors:
D. Danino, Haifa, Israel
D. Harries, Bethesda, USA
S. Wrenn, Philadelphia, USA

pp. 1–153 · December 2004
ISBN 3-527-31322-2

Macromolecular Symposia publishes lectures given at international symposia and is issued irregularly, with normally 14 volumes published per year. For each symposium volume, an Editor is appointed. The articles are peer-reviewed. The journal is produced by photo-offset lithography directly from the authors' typescripts.
Further information for authors can be found at http://www.ms-journal.de
Suggestions or proposals for conferences or symposia to be covered in this series should also be sent to the Editorial office (E-mail: macro-symp@wiley-vch.de).

Macromolecular Symposia:
Annual subscription rates 2005
Macromolecular Full Package: including Macromolecular Chemistry & Physics (24 issues), Macromolecular Rapid Communications (24), Macromolecular Bioscience (12), Macromolecular Theory & Simulations (9), Macromolecular Materials and Engineering (12), Macromolecular Symposia (14):

Europe	Euro	7,088 / 7,797
Switzerland	Sfr	12,448 / 13,693
All other areas	US$	8,898 / 9,788

print only **or** electronic only / print **and** electronic

Postage and handling charges included. All Wiley-VCH prices are exclusive of VAT. Prices are subject to change.

Single issues and back copies are available. Please ask for details at: service@wiley-vch.de

Orders may be placed through your bookseller or directly at the publishers:
WILEY-VCH Verlag GmbH & Co. KGaA, P.O. Box 10 11 61, 69451 Weinheim, Germany, Tel. +49 (0) 62 01/6 06-400, Fax +49 (0) 62 01/60 61 84, E-mail: service@wiley-vch.de

For **copyright permission** please contact Claudia Jerke at:
Fax: +49 (0) 62 01/6 06-332, E-mail: cjerke@wiley-vch.de

For USA and Canada: Macromolecular Symposia (ISSN 1022-1360) is published with 14 volumes per year by WILEY-VCH Verlag GmbH & Co. KGaA, Boschstr. 12, 69451 Weinheim, Germany. Air freight and mailing in the USA by Publications Expediting Inc., 200 Meacham Ave., Elmont, NY 11003, USA. Application to mail at Periodicals Postage rate is pending at Jamaica, NY 11431, USA. POSTMASTER please send address changes to: Macromolecular Symposia, c/o Wiley-VCH, III River Street, Hoboken, NJ 07030, USA.

Printing: Strauss Offsetdruck, Mörlenbach. Binding: J. Schäffer, Grünstadt

Macromolecular Symposia

Articles published on the web will appear several weeks before the print edition. They are available through:

www.ms-journal.de

www.interscience.wiley.com

Biocolloids: American Chemical Society Meeting Anaheim (USA), 2004

What's in a Biocolloid
D. Danino, D. Harries, S. Wrenn

Author Index

What's in a Biocolloid

In the complex world of biology, colloids present hope for simplicity. Sometimes, they serve as models that let us think about biological processes from the standpoint of physical forces. In cellular environments, lipids, proteins, and DNA assemble and disassemble, make and break into aggregates, and function with specific roles in the cellular machinery. While the processes that bring these macromolecules together are often directed by a flux of energy, occasionally biocolloids self-assemble, held together only by electrostatic, van der Waals, and hydration forces. The same forces that hold molecules together to make a colloid can also keep colloids apart. Such interplay can determine specific interaction.

At the April 2004 American Chemical Society national meeting in Anaheim, California, we – physical chemists, physicists, chemical engineers, and biologists – connected to discuss how the properties of colloidal particles are reflected in their biological role. The symposium covered a wide range of biological materials: from protein aggregates, such as microtubules and clathrin baskets, to multi-component lipid membranes. An emerging theme was how molecular material properties of aggregates can play a fundamental role in biological processes. Indeed, disease and cure can sometimes be shown to relate to flaws in, or change of, aggregate material and structural properties.

In a session devoted to lipid membranes, we discussed how cholesterol determines membrane properties and could be involved in maladies: gallstone formation, atherosclerosis, and cholesterol metabolic diseases. Domains and rafts that may form in cholesterol containing bilayers may be responsible for the modified membrane properties that in turn cause illness. Understanding the link between membrane composition and its underlying properties could serve as a key to cures. The new ways to think about the good and evil of cholesterol excited not only those who attended. They also attracted general attention that was highlighted in Chemical and Engineering News: "Chewing The Fat About Cholesterol – Molecule plays a key role in controlling the organization and fluidity of cell membranes" (May 3, 2004).

Learning about properties of lipids and lipid aggregates can aid in technological solutions to other biophysical problems, for example, how to engineer efficient and safe vectors for gene delivery or how to use lipid aggregates to align proteins for NMR studies.

In the session on superamolecular assemblies of proteins, we learned how self-assembly of proteins can be controlled not only from within by cellular regulation, to enable endocytosis and trafficking, but also from the outside by cancer drugs that control the assembly and disassembly of microtubule filaments.

The meeting focused people from diverse scientific backgrounds with common interests. We hope the printed version will be as thought-provoking as the original symposium.

Acknowledgements
The editors wish to thank Avanti Polar Lipids, Inc. and the ACS Division of Colloid and Surface Chemistry for their financial support of the Biocolloids symposium.

D. Danino
D. Harries
S. Wrenn

Assembly of Clathrin Baskets

Ralph Nossal

Laboratory of Integrative and Medical Biophysics, NICHD, National Institutes of Health, Bethesda, MD 20892, USA
E-mail: nossalr@mail.nih.gov

Summary: The creation and internalization of small, protein-coated vesicles is a central factor in the uptake of materials from the surfaces of eucaryotic cells through a process known as receptor mediated endocytosis. Under appropriate in vitro conditions, the principal coat component, viz., clathrin, which is found in the cell as a trimer joined at a common hub, assembles into polyhedral 'baskets' ('cages') containing pentagonal and hexagonal facets. These reconstituted cages have the appearance of miniscule soccer-ball like structures. Their sizes vary over a finite range whose limits depend on the presence or absence of ancillary proteins ('assembly proteins') that are known to increase the tendency for the baskets to form. By fitting data on basket size distributions to simple energetic (thermodynamic) models, one is able to estimate mechanical properties of the clathrin constituents of the baskets and infer the role of assembly proteins in strengthening interactions between the clathrin components of the struts that constitute the edges of the baskets.

Keywords: clathrin; elasticity; endocytosis; self-assembly; triskelia

Introduction

Clathrin is the primary component of the protein coats of transport vesicles involved in receptor-mediated endocytosis ("RME") and various other trafficking activities that occur in eucaryotic cells. In particular, animal cells utilize RME processes to regulate cell surface components [1]. An important example where RME plays a critical role is ErbB signaling involved in cell growth and proliferation. There are at least four members of the ErbB receptor family that form homo- or hetero-dimers which then respond differentially to growth factor ligands, depending on the particular paired combination of receptors [2]. For example, the binding of epidermal growth factor (EGF) ligands to ErbB1 dimers leads to internalization of ligand-receptor complex, but the formation of ErbB2–ErbB1 heterodimers frustrates ErbB1 uptake upon ligand binding, resulting in aberrant signaling and heightened cell growth [3,4]. Overexpression of ErbB receptors is involved in many cancers [3]; in particular, high expression of ErbB2 is a signature of certain forms of breast cancer.

 DOI: 10.1002/masy.200550101

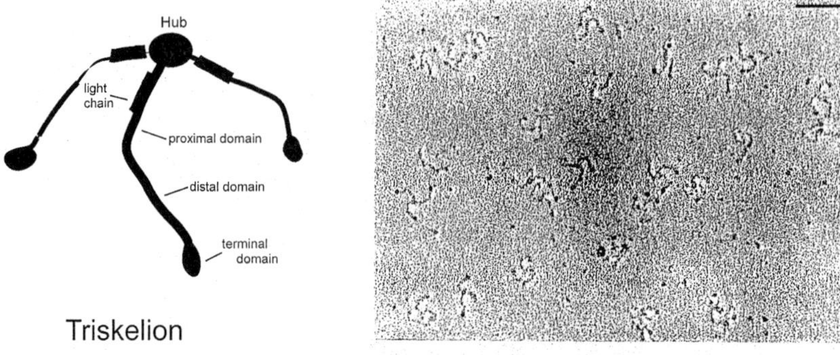

Fig. 1. Cartoon of clathrin triskelion, showing light chains joined to heavy chains near the center hub. (Electron micrograph images adapted from Ref. [10]).

When isolated from cell preparations, clathrin typically is found as a heteropolymer composed of clathrin heavy chains (CHC) of MW ca. 190 kDa and clathrin light chains (CLC) of MW ca. 25-35 kDa [5]. The native, mesoscopic, structure of clathrin is that of a three legged complex known as a "triskelion" that contains three clathrin heterodimers (each composed of a single CHC and single bound CLC), joined at a common hub [6]. The dimensions of the CHC are approximately 2.4–2.8 nm in cross-section and 50 nm when fully extended [7]. The CHC has a tendency to bend midway along its length [6-8] and the lowest energy configuration of the triskelion thus is presumed to be a pinwheel-like structure (see Fig. 1), as evidenced by electron microscopy [9,10]. Under appropriate conditions the triskelia form cage-like supramolecular structures having diameters of the order of 100 nm [9,11] and topologies similar to that of the seams of soccer balls [12]. In this form the cages contain 12 pentagonal faces and an even – but variable – number of hexagonal facets which, for energetic reasons, probably appear in multiples of 4.

The formation of plasma-membrane-derived endocytic vesicles that are involved in receptor mediated endocytosis requires several steps, the exact sequence of which has not yet been established in detail. The canonical scheme involves the binding of ligands to transmembrane receptors, the binding of activated receptors to "adaptor" proteins that move from the cytoplasm to the membrane and then recruit [13] (or perhaps are recruited by [14]) clathrin triskelia to form a coated pit, the growth and invagination of a pit to form a vesicle [15], the closure of the vesicle and its fission from the cell surface, and the subsequent movement of the vesicle to the interior of the cell where it releases cargo. The universal signature of these endocytic vesicles is a polyhedral clathrin lattice that has the appearance and approximate dimensions of reconstituted clathrin cages. Other components of the coats may be cell or organelle specific.

It is not known how the initial patch of coated membrane invaginates to form the nascent vesicle, but proteins that simultaneously bind to clathrin and certain membrane lipids (mainly phosphoinositides) – thereby possibly bending the membranes – are known to exist [16,17]. Specific proteins – notably dynamin – seem to be involved in pinching of the membrane at the neck where the vesicle detaches from the plasma membrane [18]. Also, proteins that facilitate rearrangement of the clathrin lattice near the neck may be involved [12,19,20]. (Many clathrin-associated proteins that either are integral constituents of vesicle coats, or interact with vesicles transiently during or after their formation, are discussed in a recent review [21].) In addition to being the principal coat component of receptor-linked, plasma-membrane-derived vesicles, clathrin also is found in the coats of certain tubulovesiclar bodies that emanate from the *trans*-golgi network in the interior of the cells. Similarly complex schemes, involving different coat proteins (e.g., COPI or COPII proteins instead of clathrin), give rise to other tubulo-vesicular structures involved in intracellular trafficking [22].

Given the intricacy of this system, one might at first think it impossible to ask questions of a physical nature about the processes occurring during formation of coated vesicles. Although present as isolated triskelia in vitro at pH 7.0, clathrin, alone, spontaneously assembles at low pH ($pH \leq 6.5$) to form closed cages ("baskets"). Such cages also can be assembled in the presence of "adaptors" [23], in which case they are smaller and of more uniform size [24] (see Fig. 2). These assembly factors induce basket formation even at physiological pH ($pH \approx 7.0$). The data shown here were obtained when using an imprecisely characterized fraction of "assembly proteins" (APs), which probably mostly consisted of the AP-2 adaptor complex. However, the behavior shown in Fig. 2 is typical for assembly with adaptors of various kinds [25,26]. It is of interest to note, also, that when native coated vesicles are harvested, one seemingly finds variation in mean vesicle size that depends on the source of the specimens [27].

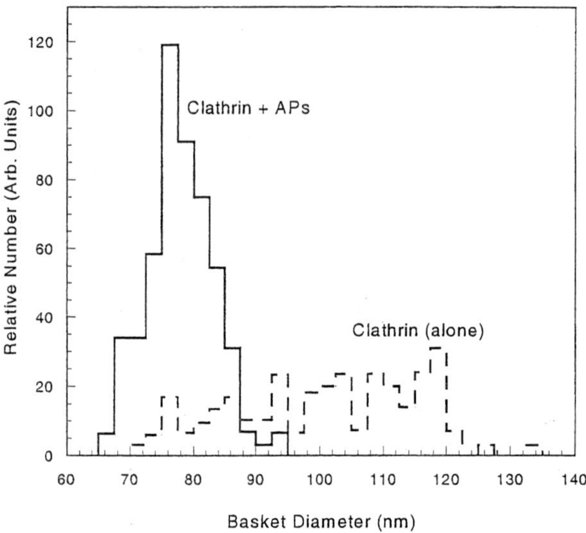

Fig. 2. Size distributions of baskets assembled in the presence (solid line) and absence (dotted line) of APs (adapted from [24]).

Inferences from Theory

Why is there dispersion in the sizes of reconstituted baskets? Clathrin cages are closed biological polymers, other examples being virus capsids [28], peptide-conjugated tubulin rings [29] and microtubules [30] (whose lengths vary widely, but whose cross-sections are closed tubes). That the baskets are closed and of finite size might be explained by the shapes of the building blocks – i.e., the triskelia. To assemble into baskets, the triskelia might have a natural "pucker." That is, they would be capable of assuming a posture where, if the ends of the legs were touching a plane, the central hub (Fig. 1) would be elevated about the surface. Indeed, such form has been inferred from carefully prepared electron microscopy images [31]. However, it is difficult to quantify the pucker from such images and, in any event, *solution* structures cannot be determined in this way. Confirmation of the existence of puckered shape in solution results from the application of static and dynamic light scattering [32]. The notion of "intrinsic pucker" of *flexible* triskelia can be used to rationalize the dispersion in basket sizes.

A simple way to characterize the energy of such a puckered entity at the mesoscopic level is

$$g^{dist} = \kappa f(\xi - \xi_0) \qquad (1)$$

where ξ is the "curvature" of the structure (integrated over the entire triskelion) and ξ_0 is the curvature of the lowest energy shape of the triskelion. κ plays the role of a rigidity modulus (or Hooke's Law constant) and, in general, $f(x \rightarrow 0) = 0$; $f(x \rightarrow \text{large}) \rightarrow$ *inf.* Since we are interested in using an expression of this nature to describe the energy of clathrin cages of "size" N, it is useful to index ξ_0 to N_0, the cage of lowest distortion energy. That is, in order to assemble a basket containing N triskelia, some triskelia need to be expanded (opened) or contracted (squeezed), or stretched or twisted (or all of these). Not enough is known about the detailed tertiary structure of a triskelion (MW ~ 650 kDa) to enable us to calculate this complex dependence which, additionally, probably depends on the position of a triskelion in an assembled cage.

To first approximation we lump these factors together, take κ to be a phenomenological constant, and consider ξ to vary inversely with N (so that f, in effect, is a function of the difference between the average solid angle subtended by a triskelion in a cage of size N and the solid angle subtended in the cage of lowest distortion energy). In this case the first term in an expansion of f (x) is $g^{dist} = \kappa (1/N - 1/N_0)^2$, which diverges when $N \ll N_0$ and tends to zero when N is close to N_0. This expression clearly fails to represent the "large N" distortion energy of a typical triskelion because $g^{dist} \rightarrow$ *constant* as N gets very large. But, the total distortion energy of a basket varies as $G_N^{dist} \sim Ng^{dist}$, so even in this approximation the total distortion energy involved in assembling a basket diverges, as expected, when the basket gets very large. Of course, baskets would not form unless there were some net favorable interaction energy between triskelia which, to first order, can be taken as $G_N^{stab} = -bN$, where b is a positive constant containing both enthalpic and entropic components. When these terms are combined, we obtain

$$G_N = aN (1/N - 1/N_0)^2 - bN \qquad (2)$$

where we have replaced the general term κ with the symbol a.

The expression given by Eq.(2) contains only three parameters: rigidity coefficient, a; the basal shape parameter N_0; and a term signifying the inter-leg interaction energy, b. Upon assuming that the baskets assemble by a thermodynamically reversible process, it follows that the equilibrium distribution of basket sizes is given by $P_N \sim \exp(-G_N/k_BT)$, which can be used to obtain expressions that relate the energy parameters {a, N_0, b} to the width of the distribution ($W_{1/2}$), the value of N where the distribution is maximal (N*), and to the values of N where P_N effectively becomes zero [33]. These can be used to fit the basket assembly data shown in Fig. 2, thereby providing estimates of various energies associated with cage assembly. Good fits are obtained [33], and one finds, for example, that under conditions used for in vitro assembly, the net energies of adding or removing triskelions to "growing" baskets are of the order of k_BT. This result agrees with observations that significant nucleotide hydrolysis apparently does not take place during coated vesicle (CV) formation, other than GTP hydrolysis associated with freeing the CVs from the plasma membrane (although ATP hydrolysis is needed to uncoat the vesicles). One also finds that the free energy change occurring when baskets form in the presence of assembly proteins ("APs") is greater than that occurring without APs [33]. It can be shown that the critical clathrin concentration for assembly, C_c, depends on the free energy change in an inverse manner. That is, the deeper is the "energy well" associated with assembly, the lower is C_c and more stable the cages that are formed ([34]), which is why the presence of APs favors basket formation.

By fitting data to this model, one also obtains an estimate of an effective "rigidity modulus" associated with distorting the triskelia. Is such assessment consistent with estimates obtained by other means? Due to its small size, it is difficult, if not impossible, to directly measure the response of a triskelion to an applied force. The only previously published estimate of triskelion rigidity relates the shape fluctuations of triskelion legs, as inferred from electron microscope images, to the flexural rigidity, EI, of the legs [8]. In that analysis the images of triskelions placed on grids by glycerol nebularization were assumed to be reasonably accurate representations of triskelion shapes in solution. The Boltzmann equation and a Hooke's Law description relating the bending energy of a thin wire to differences between stressed and relaxed shapes were used, yielding expressions for variances of Fourier mode amplitudes of triskelial shape in terms of the mode number, the length of the triskelion leg, and EI. Decomposition of the fluctuating shapes into normal modes thus yielded a value of $EI_{clath} \approx 35$ k_BT-nm for an individual leg that makes intuitive sense, in that EI_{clath} in this case is intermediate between that of cytoskeletal elements like actin filaments and microtubules, and that of soft network components such as elastin. Additionally, after making plausible assumptions about the cross sections of the composite strut linking vertices in a basket, a value of E $\approx$560 k_BT-nm was obtained for the flexural rigidity of the strut. Then, the energy necessary to bend a patch of clathrin lattice was compared with that necessary to bend plasma membrane of equivalent area. The magnitudes of these quantities were found to be similar, suggesting that the clathrin lattice can assist in the in vivo budding process but, by itself, is unlikely to be able to cause a patch of membrane to form the highly curved structures characteristic of coated vesicles.

Given the number of assumptions that enter into the analysis described in the foregoing paragraph, it is desirable to obtain another estimate for the flexural rigidity of a basket strut, EI_{strut}. By assuming that ξ in Eq.(1) represents the average *curvature* of the strut (some "lumped" characteristic of triskelion shape), an expression analogous to that of Eq.(2) can be obtained [34], viz.,

$$G_N = hN\,(1/N^{1/2} - 1/n^{1/2})^2 - bN\,, \qquad (3)$$

where h is directly proportional to EI_{strut}. Upon using this expression to reduce the data shown in Fig. 2 we find, remarkably, that the value of $EI_{strut} \approx 600$ k_BT-nm for baskets assembled in the presence of APs is in close agreement with that obtained from the earlier analysis [8] when it was assumed that the triskelion legs do not move with respect to each other within a composite strut. Moreover, the value of EI_{strut} for baskets formed in the *absence* of APs is an order of magnitude lower than in the presence of APs, but several times greater than that of an individual triskelion leg. Taken together, these results suggest that a mechanical role for APs is to prevent slippage of legs with respect to one another when a basket is stressed [manuscript in preparation].

In summary, these simple thermodynamic/energetic models rationalize the finite size-range of reconstituted cages. The use of such models to analyze basket assembly data confirms that basket growth can involve lattice rearrangements driven only by thermal mechanisms. It also provides numerical estimates of clathrin bending rigidity that agree with values inferred by analyzing electron micrographs of triskelial shapes. The models enable us to infer ways that molecular variables affect basket stability and, in particular, confirm (and provide numerical estimates of) the influence of adaptor complexes (i.e., assembly proteins) on inter-leg stabilizations. Similar models can be adapted to investigate the conditions affecting the critical concentration for basket assembly, and to extend the analysis to considerations of compound vesicles that contain membrane and/or cargo.

[1] H. Lodish, D. Baltimore, A. Berk, S. L. Zipursky, P. Matsudaira, J. Darnell, *"Molecular Cell Biology"*, 3rd ed., Scientific American Books, New York, 1995, p.722
[2] G. Gur, Y. Yarden, *Nat. Biotech.* **2004**, *22*, 169.
[3] Y. Yarden, M. X. Sliwkowski, *Nat. Rev. Mol. Cell Biol.* **2001**, *2*,127.
[4] D .S. Lidke, P. Nagy, R. Heintzmann, D. J. Arndt-Jovin, J. N. Post, H. E. Grecco, E. A. Jares-Erijman, T. M. Jovin, *Nat. Biotech.* **2004**, *22*, 198.
[5] F. M. Brodsky, C-Y.Chen, C. Knuehl, M. C. Towler, D. E. Wakeham, *Annu. Rev. Dev. Biol.* **2001**, *17*, 517.
[6] T. Kirchhausen, *Annu. Rev. Biochem.* **2000**, *69*, 699.
[7] D. E. Wakeham, J. A. Ybe, F. M. Brodsky, P. K. Hwang, *Traffic* **2000**, *1*, 393.
[8] A. J. Jin, R. Nossal, *Biophys. J.* **2000**, *78*, 1183.
[9] E. Ungewickell, D. Branton, *Nature* **1981**, *289*, 420.
[10] E. Kocsis, B. L. Trus, C. J. Steer, M. E. Bisher, A. C. Steven, *J. Struct. Biol.* **1991**, *107*, 6.
[11] C. J. Smith, N. Grigorieff, B. M. F. Pearse, *EMBO J.* **1998**, *17*, 4943.
[12] A. J. Jin, R. Nossal, *Biophys. J.* **1993**, *65*, 1523.
[13] C. M. Brown, N. O. Peterson, *Biochem. Cell Biol.* **1999**, *77*, 439.
[14] J. Rappoport, S. Simon, A. Benmerah, *Traffic* **2004**, *5*, 327.
[15] M. M. Perry, A. B. Gilbert, *J. Cell Sci.* **1979**, *39*, 257.
[16] M. G. J. Ford, I. G. Mills, B. J. Peter, Y. Vallis, G. J. K. Praefcke, P. R. Evans, H. T. McMahon, *Nature* **2002**, *419*, 361.
[17] B. Habermann, *EMBO Rep.* **2004**, *5*, 250.
[18] D. Danino, J. E. Hinshaw, *Curr. Opin. Cell Biol.* **2001**, *13*, 454.
[19] X. Wu, X. Zhao, L. Baylor, S. Kaushal, E. Eisenberg, L. E. Greene, *J. Cell Biol.* **2001**, *155*, 291.
[20] S. L. Newmyer, A. Christensen, S. Sever, *Dev. Cell* **2003**, *4*, 929.
[21] E. M. Lafer, *Traffic* **2002**, *3*, 513.
[22] B. Alberts, A. Johnson, J. Lewis, M. Raff, K. Roberts, P. Walter, *"Molecular Biology of the Cell"*, 4th ed., Garland Science, New York 2002, p.713.
[23] J. Keen, *Annu. Rev. Biochem.* **1990**, *59*, 415.
[24] S. Zaremba, J. H. Keen, *J. Cell Biol.* **1983**, *97*, 1339.
[25] R. Lindner, E. Ungewickell, *J. Biol. Chem.* **1992**, 267, 16567.
[26] M. L. Nonet, A. M. Holgado, F. Brewer, C. J. Serpe, B. A. Norbeck, J. Holleran, L. Wei, E. Hartweig, E. M. Jorgensen, A. Alfonso, *Mol. Biol. Cell* **1999**, *10*, 2343.
[27] J. Heuser, T. Kirchhausen, *J. Ultrastruct. Res.* **1985**, *92*, 1.
[28] A. Zlotnick, *Virology* **2002**, *315*, 269.
[29] N. R.Watts, N. Cheng, W. West, A. C. Steven, D. Sackett, *Biochemistry* **2002**, *41*, 12662.
[30] B. Alberts, A. Johnson, J. Lewis, M. Raff, K. Roberts, P. Walter, *"Molecular Biology of the Cell"*, 4th ed., Garland Science, New York 2002, p.908.
[31] T. Kirchhausen, S. C. Harrison, J. Heuser, *J. Ultrastruct. Mol. Struct. Res.* **1986**, *94*, 199.
[32] M. Ferguson, H. Boukari, D. Sackett, R. Nossal, *Biophys. J.* **2004**, *86*, 234a.
[33] R. Nossal, *Traffic* **2001**, *2*, 138.
[34] R. Nossal, M. Muthukumar, *Biophys. J.* **2003**, *84*, 209a.

Ring Polymers of Tubulin Induced by Binding of Natural Antimitotic Peptides

Dan L. Sackett

Laboratory of Integrative and Medical Biophysics, National Institute of Child Health and Human Development, National Institutes of Health, Bldg 9, Rm 1E129, 9000 Rockville Pike, Bethesda, MD, 20892, USA
E-mail: sackettd@mail.nih.gov

Summary: A number of natural products with potent antimitotic activity are peptides and depsipeptides that bind to tubulin, provoke depolymerization of microtubules, and induce formation of single layer rings of tubulin dimers. These peptides are all hydrophobic and small relative to tubulin (3-5 amino acid residues), yet induce rings polymers with properties that can differ significantly in size and self-association. In addition, these compounds exhibit potent cytotoxicity that varies by several hundred fold from one compound to another. Cryptophycin induces unusually homogeneous rings, composed of eight tubulin dimers, that are stable to dilution at least to nanomolar tubulin concentrations.

Keywords: biopolymers; cryptophycin; microtubules; ring polymers; supramolecular structures

Introduction

Microtubules (MT) are the largest of the three polymer fiber systems that together comprise the cytoskeleton of eukaryotic cells. MT are dynamic noncovalent polymers composed of a structural subunit, the heterodimer tubulin, and a large number of associated proteins which regulate assembly, promote directed movement along the MT surface, and form signaling complexes on MT. Assemblies of MT provide the cell with polarity (such as inside – outside directionality), and regulate directed intracellular movement. Notably, arrays of MT comprise the mitotic spindle, the molecular machine responsible for accurate separation of the two sets of chromosomes at cell division.

This highly complex and dynamic structure is very sensitive to alteration or disruption of MT dynamics or function. A large number of small molecules are known that are capable of doing this, and are collectively known as mitotic poisons. The vast majority of these exert their effect by binding to and altering tubulin, and many of these are, or are derived

 DOI: 10.1002/masy.200550102

from, natural products. The large number of these found in nature suggests that MT and mitosis are particularly vulnerable targets to which organisms have evolved chemical attacks, presumably as a defense from predators or as a way to achieve advantage over competitors.

Antimicrotubule Agents

Antimitotic agents targeting MT fall into several classes, based on their non-overlapping binding sites on the tubulin dimer. The classic mitotic poison colchicine defines one of the major binding sites on tubulin, and many natural, semisynthetic, and synthetic compounds are known that bind to this site. A second site is defined by the binding of vinblastine and related vinca alkaloids. Many natural compounds unrelated to vinblastine nonetheless have binding sites that appear to at least overlap with that of vinblastine. These compounds tend to be larger in molecular mass than those binding to the colchicine site. A third site is defined by the binding of taxol, and a small but growing group of compounds are known whose binding is at the taxol site. An additional "site" is the cysteine sulfhydryl groups exposed on the tubulin dimer.

Table 1. Properties of molecules binding to the three major drug binding sites on tubulin. These are grouped by their effect on MT, and then by their binding site. Colchicine site agents tend to be smaller molecules than vinca site binders.

MT effect	Depolymerizing		Hyperstablizing
Binding site on:	Tubulin dimer		Tubulin polymer (MT)
Binding site:	Colchicine	Vinca	Taxol
Examples:	Colchicine Podophyllotoxin Combretastatin Nocodazole	Vinblastine Vincristine Maytansine Rhizoxin	Taxol Epothilone Discodermolide
Average MW, Da Example (MW), Da	~ 370 Colchicine (400)	~750 Vinblastine (800)	~650 Taxol (850)

Binding at all of these except for the taxol site results in inhibition of MT polymerization, while occupation of the taxol site results in hyperstabilization of the MT and inhibition of depolymerization. Both polymerization and depolymerization are required for the normal dynamic functioning of MT and indeed all of these agents have been shown to alter dynamics at lower concentrations than those required for effects on total polymer mass[1]. The properties of agents binding to these binding sites are summarized in Table 1, with a few examples of molecules in each class.

Vinca Domain Peptide Agents

In addition to vinca site agents such as those listed in Table 1, a subclass of compounds is composed of peptides and depsipeptides. These inhibit the binding of vinblastine to tubulin, as required for a vinca site agent, but their inhibition is not competitive, leading to the suggestion that their binding site(s) are overlapping but non-identical with that for vinblastine. This has given rise to the name "vinca domain" for the collection of binding sites overlapping with that for vinblastine[2].

Three interesting examples of the peptide agents binding in the vinca domain are hemiasterlin, cryptophycin 1, and dolastatin 10 [3]. These are, respectively, a tripeptide, a cyclic four residue depsipeptide, and a linear pentapeptide, and in all three the amino acid residues are highly modified. All three are hydrophobic small peptides, and all are mutually competitive inhibitors of binding. Their structures are given in Figure 1. Like all vinca agents, these strongly destabilize MT and disfavor polymerization. Unlike many other destabilizing agents, these compounds do not simply prevent tubulin from polymerizing into MT, the tubulin remaining as dimers. Rather, these compounds induce tubulin to form alternative oligomeric structures[4]. Vinblastine also does this, inducing the formation of linear spirals of indefinite length. These compounds also induce formation of oligomers, but these are closed oligomers, specifically rings of defined size.

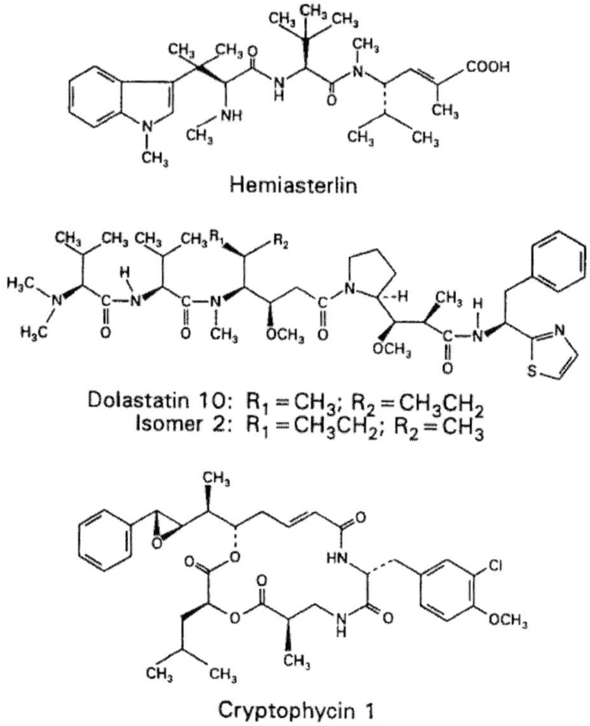

Hemiasterlin

Dolastatin 10: $R_1 = CH_3$; $R_2 = CH_3CH_2$
Isomer 2: $R_1 = CH_3CH_2$; $R_2 = CH_3$

Cryptophycin 1

Fig. 1. Structures of the three peptide antimitotic agents.

Structure and Properties of the Peptide-Induced Rings

Interestingly, the rings formed by the three peptides are similar but distinct. The properties of the rings were revealed by electron microscopy (yielding the ring diameter, and, by STEM analysis, the mass), a combination of sedimentation velocity (to yield sedimentation coefficients) and dynamic light scattering (to yield diffusion coefficients), the latter combination providing the ring mass by the Svedberg equation[5,6]. The results are presented in Table 2.

The rings are all formed by a single linear chain of dimers, unlike other known tubulin rings, e.g. the tubulin GDP-Mg rings which are double unequal nested rings[7], or the Rev-tubulin rings[8], which are double equal stacked rings. While the peptide rings are all single filament rings, their diameter, and hence curvature, are not the same.

Table 2. Properties of peptide-induced tubulin ring polymers.[5,6]

Drug	Diameter EM, nm	Sedimentation Coeffifient (s)	Diffusioin Coefficient $(cm^2sec^{-1})x10^7$	Ring Mass kDa	Number of Tubulin Dimers
Cryptophycin	27 ± 1	16	1.8	820	8
Hemiasterlin, Dolastatin	42 ± 2	20	1.3	1400	14

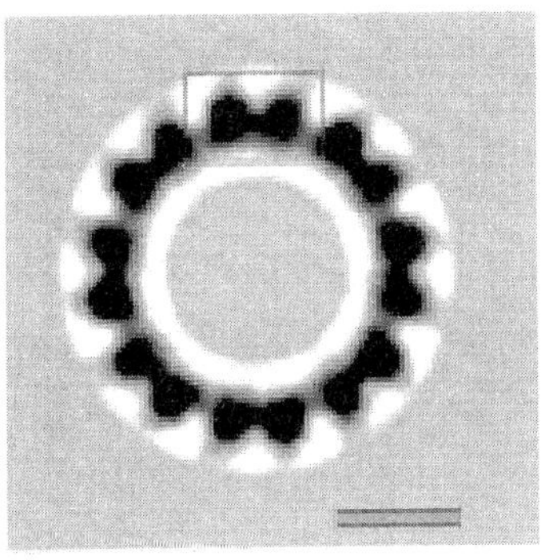

Fig. 2. The cryptophycin-tubulin ring, reconstructed from cryoelectron microscopy[5]. An individual tubulin dimer is outlined in the box; the β-subunit is on the left. The bar = 10 nm.

The cryptophycin rings are the smallest tubulin rings reported[5,9] , composed of only 8 dimers. We find the same size rings with cryptophycin-1, the natural product, or cryptophycin-52, an analog in clinical trials. The rings are also very homogeneous, with both compounds. This uniformity has facilitated structural analysis of these rings by cryoelectron microscopy and image reconstruction[5]. The resulting image, Figure 2, shows the 8-dimer ring of uniform polarity (no reversals of apparent handedness), with a clearly larger inter-dimer spacing than intra-dimer spacing. There are two points of curvature in each dimer: a larger angle (32°) at the inter-dimer contact and a smaller angle (13°) at the intra-dimer contact. Additionally, the normally smooth surface on the outside of the MT is now on the interior of the ring, and the scalloped morphology normally on the interior of the MT is now evident on the outside of the ring. Clearly cryptophycin has had the effect of causing a normal MT protofilament to curl into a ring, with the dimer surface normally exposed on the outside of the MT now confined in the center of the ring. Protofilaments need not pre-exist for ring formation; addition of cryptophycin (or dolastain or hemiasterlin) to tubulin dimers results in the same ring structures as if added to MT. Dolastatin and hemiasterlin rings are the same diameter and are both larger than cryptophycin rings, composed mostly of 14 dimers, and are more heterogeneous in size. While these two ring types have the same morphology, they differ in that the dolastatin rings self-associate readily, causing significant increases in solution turbidity due to formation of structures composed of stacked rings that grow large enough to settle under gravity[3,10]. Hemiasterlin rings, like cryptophycin rings, do not self-associate under the same conditions, and the solutions remain water-clear.

The rings also differ in their stability to dilution. Fluorescence correlation spectroscopy (FCS) allows the diffusion coefficient of the rings to be estimated at a very wide range of concentrations. This analysis showed that cryptophycin rings exhibit no measurable depolymerization even when diluted to 1 nM total tubulin. Dolastatin rings depolymerized measurably below 10 nM, while hemiasterlin rings were significantly depolymerized at 100 nM [11]. This is the same order as the cytotoxicity of these compounds.

A comparison of the cytotoxicities of these compounds in presented in Table 3. This table also compares these compounds with vinblastine and taxol, demonstrating the significantly increased potency of these compounds (as much as 500-fold compared to taxol).

Table 3. Cytotoxicity of peptides on different cell lines. The IC_{50} (in nM) for a 4-day incubation wih drug is shown for ovarian (1A9), prostate (PC3), and breast (MCF7) carcinoma, and for lymphoma (Ca46) cell lines.[11]

Drug	1A9	PC3	MCF7	Ca46
Cryptophycin	0.01	0.016	0.01	0.012
Dolastatin	0.05	0.08	0.04	0.06
Hemiasterlin	0.24	0.6	0.3	0.2
Vinblastine	1.1	1.9	1	1.1
Taxol	5	8	6	6

These results suggest that the molecular properties of the rings may reflect parameters of the peptide-tubulin interaction that are significant for cell toxicity, whether or not the rings actually form inside treated cells. Additionally, it highlights the ability of these small molecules to induce formation of nanoscopic structures of high reproducibility. Finally, the modular synthesis of these compounds[2] offers the possibility of "tuning" the interaction with tubulin by alteration of the structure of the peptides.

It is interesting to consider several unresolved points. Why do the peptides induce rings of differing curvature? Why do the rings differ in stability to dilution? Why do dolastatin rings aggregate but the others do not?

Computational docking studies of cryptophycin 52[9] and cryptophycin 1, dolastatin 10, and hemiasterlin[12] have indicated that the binding site for these compounds is on ß-tubulin adjacent to the exchangeable GTP site, in the inter-dimer interface. The site is somewhat to the "side" of the dimer and hence is involved in lateral interactions with the neighboring protofilament in a MT[9]. Nonetheless, occupancy of this site requires an upward displacement of helix H10 of the α-subunit of the next dimer[12], and this displacement may be the origin of the curvature observed at the inter-dimer contact[5]. It will be interesting to see if the displacement required by docking of cryptophycin is more than that required for hemiasterlin and dolastatin, as would be necessary to produce smaller rings.

The rings differ in stability to dilution. One explanation of this would be that this stability simply reflects the tightness of binding. The binding site is located at the contact site between neighboring dimers and is predicted to feature contacts between the peptide and both sides of the inter-dimer junction[9,12]. Therefore an increased binding affinity would plausibly result in tighter association of neighboring dimers and hence a more stable ring polymer. However, the binding affinities deduced from computational docking, as well as available experimental values, indicate the wrong trend for this explanation[12]. Hemiasterlin binding affinity is more than 10 fold higher than that for cryptophycin, yet the cryptophycin rings are far more stable than those formed with hemiasterlin. Hence, the explanation for stability differences must be more subtle.

The rings differ in association behavior, and the reason is not clear. Dolastatin rings aggregate readily while the other rings do not. At room temperature and pH 7, dolastatin rings associate to form structures that generate significant turbidity. These structures become large enough that they settle out of solution under 1 g. Under similar conditions, the cryptophycin and hemiasterlin rings do not aggregate, as indicated by the absence of turbidity, and by unchanged behavior in FCS. It seems remarkable that these rings are so different, since the peptides are all small hydrophobic peptides that bind to the same site on tubulin, the tubulin is the same in all cases, and the buffer is also the same. The rings that are produced have the same morphology (i.e. the dimer surface exposed on the outside of the MT is on the inside of the ring), and are even the same diameter in the case of hemiasterlin and dolastatin. A related puzzle is that addition of substoichiometric amounts of cryptophycin or hemiasterlin to a mixture of dolastatin and tubulin poisons the self-association of the resulting rings[3]. In the case of cryptophycin addition, this might be rationalized based on the smaller size of the cryptophycin rings: perhaps the association requires rings of similar size. But this would fail to account for the behavior of the hemiasterlin – dolastatin rings, which are identical in size. Thus there must be some subtle difference in conformation of the rings. Since small angle neutron scattering reveals that the dolastatin rings stack like coins to form these structures[10], the differences presumable are to be found in the surfaces of the rings that contact in these stacks. These dimer surfaces are those that are involved in lateral contacts in MT. Further research will be required to reveal the nature of these subtle changes as well as the details that account for the size and stability differences.

[1] M.A. Jordan, *Curr. Med. Chem. Anti-cancer Agents* **2002**, 2,1.
[2] E. Hamel, D.G.Covell, *Curr. Med. Chem.Anti-cancer Agents* **2002**, 2, 19.
[3] R. Bai, N.A. Durso, D.L. Sackett, E. Hamel, *Biochemistry* **1999**, 38, 14302.
[4] D. Panda, R.H. Himes, R.H. Moore, L. Wilson, M.A. Jordan, *Biochemistry* **1997**, 36, 12948
[5] N.R. Watts, N. Cheng, W. West, A.C. Steven, D.L. Sackett, *Biochemistry* **2002**, 41, 12662.
[6] H. Boukari, R. Nossal, D.L. Sackett, P. Schuck, *Phys. Rev. Lett.* **2004**, 93, 098106.
[7] W.V. Nicholson, M. Lee, K.H. Downing, E. Nogales, *Cell Biochem. Biophys.* **1999**, 31, 175.
[8] N.R. Watts, D.L. Sackett, R.D. Ward, M.W. Miller, P.T. Wingfield, S.S. Stahl, A.C. Steven, *J. Cell Biol.* **2000**, 150, 349.
[9] P. Barbier, C. Gregoire, F. Devred, M. Sarrazin, V. Peyrot, *Biochemistry* **2001,** 40, 13510.
[10] H. Boukari, V. Chernomordik, S. Krueger, R. Nossal, D.L. Sackett, *Physica B*, **2004**, 350, e533.
[11] H. Boukari, R. Nossal, D.L. Sackett, *Biochemistry* **2003**, 42, 1292.
[12] A. Mitra, D. Sept, *Biochemistry*, 2004, In Press.

Studying Spatial Distributions of Influenza Hemagglutinin on the Plasma Membrane of Fibroblasts: A Work in Progress

Joshua Zimmerberg,[*1] Mukesh Kumar,[1] Anil Verma,[1] Jane Farrington,[2] Michael Roth,[3] Anne Kenworthy,[1] Samuel T. Hess[4]*

[1] Laboratory for Cellular and Molecular Biophysics, National Institute of Child Health and Human Development, National Institutes of Health, 10 Center Drive, Bethesda, MD 20892-1855, USA
E-mail: joshz@helix.nih.gov
[2] Department of Biochemistry, University of Texas Southwestern Medical Center at Dallas, 5323 Harry Hines Blvd., Dallas, TX 75390-9038, USA
[3] Departments of Molecular Physiology & Biophysics and Cell and Developmental Biology, Vanderbilt University School of Medicine, Nashville, Tennessee 37232, USA
[4] Department of Physics and Astronomy, University of Maine, Orono, ME 04469, USA

Summary: The major envelope protein of influenza virus, hemagglutinin (HA), mediates the fusion of virus to cell for infection, and can mediate cell-cell fusion. It has been studied as a "raft" protein, as it is found in detergent-resistant membranes (DRM) and trafficks apically in polarized epithelia. Moreover, the viral envelope of influenza itself is rich in sphingomyelin and cholesterol. Using both immunogold electron microscopy and fluorescence resonance energy transfer (FRET) microscopy, we are examining the distribution of HA on the surface of fibroblasts expressing wild-type HA.

Keywords: cholesterol; fibroblasts; FRET; microdomains; raft; sphingomyelin

The complexities of biomembrane structure are of great interest because membranes are involved ubiquitously in biological function. In particular, the spatial organization of membrane lipids and proteins into sub-micron sized domains enriched in cholesterol, sphingolipids, and saturated lipids ("rafts") is controversial because of the paucity of methods to visualize membrane domains directly on appropriate length scales (Anderson and Jacobson, 2002a). Detergent-insolubility at low temperatures has been used as the operational definition of "raft" constituents (Brown and London, 2000) but also provides, at best, only indirect information. Nonetheless, there is a correlation between detergent-insolubility of proteins and many biological functions: assembly and budding of HIV, Ebola, and influenza (Ali et al., 2000; Bavari et al., 2002; Nguyen and

 DOI: 10.1002/masy.200550103

Hildreth, 2000; Ono and Freed, 2001; Scheiffele et al., 1999; Zhang et al., 2000) appear to occur in or depend on microdomain lipids, and membrane fusion catalyzed by certain proteins such as influenza hemagglutinin (HA), appears to be disrupted by removal of membrane cholesterol by methyl-*β*-cyclodextrin (M*β*CD) (Sun and Whittaker, 2003). These effects are thought to result from redistribution of the HA and lipid after disruption of putative "raft" domains which presumably restrict lateral diffusion and thereby maintain a high local concentration of HA (Takeda et al., 2003).

Microdomains are also proposed to play a role in protein sorting and trafficking, cell signaling, toxin and pathogen binding, and potentially the faulty steps that lead to protein aggregation in Alzheimer's disease (Brown and London, 2000; Simons and Ikonen, 1997). Extraction of cholesterol and lipids by M*β*CD treatment has been shown to disrupt aggregation and function of microdomain-associated proteins such as Lyn kinase and its signal transduction pathway with the IgE receptor (Sheets et al., 1999). However, the unperturbed structure of membrane microdomains has not yet been visualized, nor have many of their physical properties such as lipid phase state been established in living cells at physiological temperatures. Visualization and physical characterization of such domains will almost certainly help illuminate how they perform their numerous biological functions.

The structure of membranes on the molecular level is clearly far more complex than the traditional view of the fluid mosaic model.(Simons and Ikonen, 1997; Singer and Nicolson, 1972) However, current models do not find a consensus on membrane organization (Anderson and Jacobson, 2002b; Brown and London, 2000; Edidin, 2001; Edidin, 2003; Fujiwara et al., 2002; Kwik et al., 2003; Maxfield, 2002; Sharma et al., 2004; Simons and Ikonen, 1997); how proteins and lipids orchestrate the large number of biological functions that occur in cellular membranes, on what spatial scale, and in what lipid phase various functions are performed, remains largely unclear. Distinction among these various models would provide fundamental insight into how cells and viruses organize their membranes.

The *lipid shell* model (Anderson and Jacobson, 2002b), similar to the boundary lipid model introduced in the late 1970s (Moore et al., 1978), depicts lipids preferentially associating with certain proteins in a layer of a few molecules (~7 nm) in thickness. Thus, rafts would be defined by direct interactions between proteins and their nearest protein or lipid neighbors, and would constitute a small total number of molecules. The *fluid phase* raft model depicts liquid ordered membrane domains (Brown and London, 2000; Ge et al., 1999; Ge et al., 2003; Maxfield, 2002; Munro, 2003; Rietveld and Simons, 1998) composed of lipids enriched in cholesterol and sphingomyelin which preferentially associate with each other and certain proteins (Brown and London, 2000; Ge et al., 1999). In this model, raft-associated proteins and lipids partition preferentially into raft (liquid ordered) domains (Brown and London, 2000; Harder and Simons, 1997; Rietveld and Simons, 1998). However, the proteins may not necessarily bind lipid molecules directly, in contrast with the lipid shell model. For example, it has been proposed that glycosyl-phosphatidyl-inositol (GPI)-linked proteins may be largely unclustered within domains despite preferential partitioning into those domains (Brown and London, 2000). In this model, diffusion or flow of proteins within fluid domains may occur independently of other proteins and lipids. The *solid phase* raft model depicts solid crystalline (also called gel) phase (Munro, 2003) membrane domains which consist of preferentially associating lipids, also enriched in cholesterol and sphingomyelin, in which certain proteins are preferentially soluble. However, solid phase domains are not expected to allow relative motion of domain constituents and have been largely dismissed as incompatible with cellular function (Brown and London, 2000) and with changes in lateral mobility of membrane constituents (Kenworthy et al., 2004). The *picket fence* model (Fujiwara et al., 2002; Nakada et al., 2003) depicts proteins and lipids as free to diffuse laterally within domains contained by barriers composed of stationary membrane proteins, but confined by these barriers on millisecond timescales. Traversal of the barriers ("hopping") is possible but may require multiple attempts. On the other hand, membrane heterogeneity may be maintained by continual site-dependent *endo- and exocytosis*, rather than by specific lateral interactions between membrane components(Edidin, 2003). Finally, it is possible that a combination of the above (*multiple domain models*)(Maxfield, 2002) best describes the organization of cell membranes.

Recently, we have focused our attention on testing the predictions of the fluid domain raft model for several reasons: 1. the fluid domain raft model has recently received considerable attention in studies of cell membranes and biomembrane models (Baumgart et al., 2003; Brown and London, 2000; Rietveld and Simons, 1998; Veatch and Keller, 2002; Veatch and Keller, 2003); (Rietveld and Simons, 1998) 2. the domains of HA we observe (see below) are much larger than a few times the molecular dimension, eliminating domain models which predict organization exclusively on molecular length scales 3. fluid domains are expected to have distinct, testable properties, including rounded boundaries, preferential partitioning of certain proteins and lipids into those domains, and a uniform density of constituents within the domains. We are in the process of testing the predictions of the numerous raft models described in literature, and in particular the fluid phase raft domain model, for consistency or inconsistency with our observations of the lateral organization of HA on fibroblast cell plasma membranes.

The lipid-dependent clustering of HA in cellular membranes has been implicated previously by detergent-extraction studies (Melkonian et al., 1999; Tatulian and Tamm, 2000) and electron microscopy (Takeda et al., 2003). However, mutations in HA which alter its fractionation after detergent treatment (*i.e.* DRM or "raft" association) (Melkonian et al., 1999) do not necessarily reduce fusion competence (Melikyan et al., 1997), and microdomain association does not depend on viral infection (Skibbens et al., 1989). Thus the connection between microdomain association and function could be clarified by direct one-dimensional and (especially) two-dimensional visualization of the clusters. For example, if the function of microdomain association is to assist influenza virus in exiting from an infected cell, as has been suggested in the case of vesicular stomatitis virus glycoprotein (Brown and Lyles, 2003), the fusion-associated domain could function independently. Hence, lateral clustering of HA may play a role in its infectivity at two points in its life cycle: 1. membrane fusion which mediates viral entry into the host cell, 2. concentration of HA and other viral proteins into domains which will eventually bud from the cell as new viruses. Thus, we propose that the transmembrane domain sequence of HA may have evolved to optimize the protein and lipid composition of the viral envelope, which is formed in the cell membrane prior to budding. We suggest that the mechanism of concentration of viral proteins by rafts (see above) may be widespread in enveloped viruses, since other viral glycoproteins have been shown to aggregate into 100-150 nm membrane domains that might serve as viral budding sites (Brown and Lyles, 2003; Takeda et al., 2003).

Here, electron microscopy has proven useful because it provides access to crucial nanometer length scales (Fig. 1). We are currently determining whether HA expressed on fibroblast cell plasma membranes is distributed randomly or non-randomly, to determine whether the distribution is lipid-dependent, and to determine which models of membrane "raft" organization are consistent or inconsistent with the observed distribution. Because constraint of lipid motion is also closely linked to the steps leading to fusion of influenza virus with host cells (Chernomordik et al., 1998), the question of the spatial distribution of HA on cell surfaces has many important consequences.

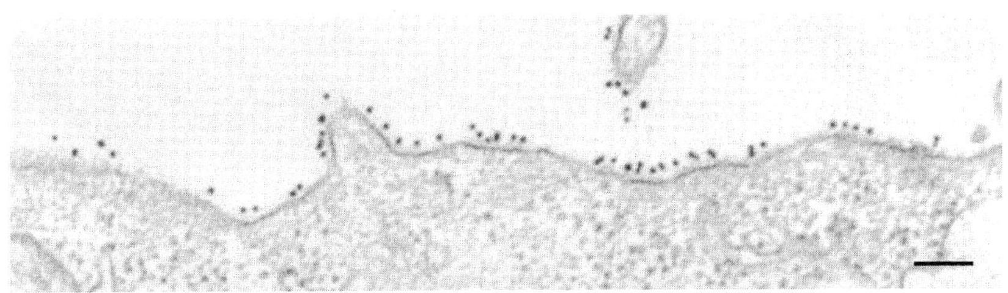

Fig. 1. HAb2 fibroblast immunostained for HA with 10nm goat anti mouse secondary antibody and then fixed. Bar = 10nm.

Quantitative methods exist for analysis of arbitrary particle distributions and can extract information about inter-particle interactions (Prior et al., 2003; Ripley, 1977; Ripley, 1979). Quantification of clustering using multiple types of statistical analysis is crucial since many factors can mislead the eye, and each statistical method will have its strengths and weaknesses. For example, a reduction in the antibody binding probability (capture ratio) can erroneously appear to suggest a reduction in clustering and significantly affect the shape and peak of the frequency distribution of nearest neighbor distances, even when the pattern of clustering has not actually changed. We are currently unifying a number of useful methods from diverse fields to probe the spatial pattern of clustering of influenza HA and its lipid-dependence. Furthermore, many of these methods are applicable to both one-dimensional (serial section) and two-dimensional (*en face*) visualizations of particles. *En face* images typically have the advantage of larger numbers of particles given the same linear dimension of the visualized membrane patch, and provide additional types of information such as biases in the angular orientation of particles and their neighbors. In particular, the methods we have chosen are relatively model-independent and allow for the possibility of either a random or a clustered membrane distribution. We hope to distinguish between membrane models which result in uniform density within domains due to preferential partitioning of molecules into those domains, and models of domains resulting from molecular aggregation by direct molecule-molecule interactions.

This approach may help determine if the measured HA distribution is consistent or inconsistent with partitioning of trimers into domains with uniform intra-domain density, which will help to explain the general inconsistency within the raft field in reported domain sizes, depending on the specific experimental method and its corresponding, accessible range of length scales (Anderson and Jacobson, 2002b). For such purposes, electron microscopy is advantageous compared to other methods which probe a narrower range of length scales, such as fluorescence resonance energy transfer (FRET) which can access from 1 to ~10 nm, or confocal microscopy, which is barely capable of resolving such domains, which are apparently smaller than <1 μm in lateral dimension in most cells at physiological temperatures. The ability of immunogold electron microscopy to probe such a wide range of length scales (~10 nm to > 1 μm) places it in an unusual category of techniques which bridge between the molecular and the microscopic regimes. The combination of immunoelectron microscopy with FRET can yield relevant information over more than three orders of magnitude in length (~10-9 m to 10-6 m). This is exactly the range of distances that will be important in unraveling the various contributions of protein-protein and protein-lipid energetics to microdomain-based augmentation of enzyme kinetics, signaling cascades, and pathogenesis by infectious agents.

[1] Ali, A., R.T. Avalos, E. Ponimaskin, and D.P. Nayak. 2000. *In* J Virol. Vol. 74. 8709-19.
[2] Anderson, R.G.W., and K. Jacobson. 2002a. Cell biology *Science*. 296:1821-1825.
[3] Anderson, R.G.W., and K. Jacobson. 2002b. *Science*. 296:1821-1825.
[4] Baumgart, T., S.T. Hess, and W.W. Webb. 2003. *Nature*. 425:821-824.
[5] Bavari, S., C.M. Bosio, E. Wiegand, G. Ruthel, A.B. Will, T.W. Geisbert, M. Hevey, C. Schmaljohn, A. Schmaljohn, and M.J. Aman. 2002 *In* J Exp Med. Vol. 195. 593-602.
[6] Brown, D.A., and E. London. 2000. 275:17221-17224.
[7] Brown, E.L., and D.S. Lyles. 2003. *Journal of Virology*. 77:3985-3992.
[8] Chernomordik, L.V., V.A. Frolov, E. Leikina, P. Bronk, and J. Zimmerberg. 1998. *Journal of Cell Biology*. 140:1369-1382.
[9] Edidin, M. 2001. 11:492-496.
[10] Edidin, M. 2003. *Nat Rev Mol Cell Biol*. 4:414-8.
[11] Fujiwara, T., K. Ritchie, H. Murakoshi, K. Jacobson, and A. Kusumi. 2002. *Journal of Cell Biology*. 157:1071-1081.
[12] Ge, M., K.A. Field, R. Aneja, D. Holowka, B. Baird, and J.H. Freed. 1999. *Biophysical Journal*. 77:925-933.
[13] Ge, M., A. Gidwani, H.A. Brown, D. Holowka, B. Baird, and J.H. Freed. 2003. *Biophys. J.* 85:1278-1288.
[14] Harder, T., and K. Simons. 1997. *Current Opinion in Cell Biology*. 9:534-542.
[15] Kenworthy AK, B.J. Nichols, C.L. Remmert, G.M. Hendrix, M. Kumar, J. Zimmerberg, and J. Lippincott-Schwartz. 2004. *J. Cell Biol.* Jun 7;165(5):735-46.
[16] Kwik, J., S. Boyle, D. Fooksman, L. Margolis, M.P. Sheetz, and M. Edidin. 2003. Membrane cholesterol, lateral mobility, and the phosphatidylinositol 4,5-bisphosphate-dependent organization of cell actin. *Proc Natl Acad Sci U S A*. 100:13964-9.
[17] Maxfield, F.R. 2002. *Current Opinion in Cell Biology*. 14:483-487.
[18] Melikyan, G.B., H. Jin, R.A. Lamb, and F.S. Cohen. 1997. *Virology*. 235:118-128.
[19] Melkonian, K.A., A.G. Ostermeyer, J.Z. Chen, M.G. Roth, and D.A. Brown. 1999. *Journal of Biological Chemistry*. 274:3910-3917.
[20] Moore, B.M., B.R. Lentz, and G. Meissner. 1978. *Biochemistry*. 17:5248-55.
[21] Munro, S. 2003. *Cell*. 115:377-388.
[22] Nakada, C., K. Ritchie, Y. Oba, M. Nakamura, Y. Hotta, R. Iino, R.S. Kasai, K. Yamaguchi, T. Fujiwara, and A. Kusumi. 2003. *Nature Cell Biology*. 5:626-U3.
[23] Nguyen, D.H., and J.E. Hildreth. 2000. *In* J Virol. Vol. 74. 3264-72.
[24] Ono, A., and E.O. Freed. 2001. *Proc Natl Acad Sci U S A*. 98:13925-30.
[25] Prior, I.A., C. Muncke, R.G. Parton, and J.F. Hancock. 2003. *Journal of Cell Biology*. 160:165-170.
[26] Rietveld, A., and K. Simons. 1998.. *Biochimica et Biophysica Acta (BBA) - Reviews on Biomembranes*. 1376:467-479.
[27] Ripley, B.D. 1977. *Journal of the Royal Statistical Society Series B-Methodological*. B39:172-192.
[28] Ripley, B.D. 1979. *Journal of the Royal Statistical Society Series B-Methodological*. 41:368-374.
[29] Scheiffele, P., A. Rietveld, T. Wilk, and K. Simons. 1999. *Journal of Biological Chemistry*. 274:2038-2044.
[30] Sharma, P., R. Varma, R.C. Sarasij, Ira, K. Gousset, G. Krishnamoorthy, M. Rao, and S. Mayor. 2004. *Cell*. 116:577-89.
[31] Sheets, E.D., D. Holowka, and B. Baird. 1999. *Journal of Cell Biology*. 145:877-887.
[32] Simons, K., and E. Ikonen. 1997. *Nature*. 387:569-572.
[33] Singer, S.J., and G.L. Nicolson. 1972. *Science*. 175:720-731.
[34] Skibbens, J.E., M.G. Roth, and K.S. Matlin. 1989. *Journal of Cell Biology*. 108:821-832.
[35] Sun, X., and G.R. Whittaker. 2003.. *In* J Virol. Vol. 77. 12543-51.
[36] Takeda, M., G.P. Leser, C.J. Russell, and R.A. Lamb. 2003. *Proc Natl Acad Sci U S A*. 100:14610-7.
[37] Tatulian, S.A., and L.K. Tamm. 2000. *Biochemistry*. 39:496-507.
[38] Veatch, S.L., and S.L. Keller. 2002. *Physical Review Letters*. 89:art. no.-268101.
[39] Veatch, S.L., and S.L. Keller. 2003. *Biophysical Journal*. 84:725-726.
[40] Zhang, J., A. Pekosz, and R.A. Lamb. 2000. *In* J Virol. Vol. 74. 4634-44.

Macromol. Symp. **2005**, *219*, 25-38

Membrane Cholesterol and the Formation of Cholesterol Domains in the Pathogenesis of Cardiovascular Disease

*Gregory M. Troup, Yi Xie, Kathleen Boesze-Battaglia, Yong Huang, Terry Kirk, Francine Hanley, Thomas N. Tulenko**

Departments of Surgery, Biochemistry & Molecular Pharmacology, Thomas Jefferson University College of Medicine, 1205 Walnut Street, Suite 605, Philadelphia, PA 19107, USA
E-mail: thomas.tulenko@jefferson.edu

Summary: At least three types of cholesterol-rich membrane domains have been described in biological membranes including cholesterol rafts, membrane caveolae and crystalline cholesterol domains,. While clear biological functions have been ascribed to both rafts and caveolae, little attention has been directed to the biological consequences of cholesterol enrichment of cell membranes and the formation of cholesterol domains. Elevated blood cholesterol levels have been shown to result in the enrichment of the cell plasma membrane with cholesterol in arterial smooth muscle cells (SMC), endothelial cells (EC) and cardiac myocytes. In the early period of cholesterol feeding (within days), the cell membrane enriches with cholesterol and membrane viscosity and membrane bilayer width increase. This latter effect severely alters membrane protein function, and recent data indicates that this induces the modulation of vascular cells (SMC and EC) to the atherosclerotic phenotype. In cardiac myocytes these membrane modifications appear to induce alterations in gene expression patterns that lead to the development of a heart failure phenotype. In addition, as the cholesterol content increases, phase separation of cholesterol occurs resulting in the formation of immiscible cholesterol domains within the membrane. These domains likely initiate nucleation of cholesterol crystals which would explain the origin of "cholesterol clefts" in atherosclerotic lesions. Taken together, these membrane alterations secondary to cholesterol enrichment constitute a "membrane lesion" which contribute to the very early pathogenic events underlying major human diseases including coronary artery disease, stroke and heart failure.

Keywords: atherosclerosis; cardiac myocytes; endothelium; heart failure; smooth muscle

Introduction

At its most basic level, the cell's plasma membrane serves the essential task of keeping the inside of the cell in and the outside of the cell out. To accomplish this nature has evolved a lipid bilayer envelope to surround the cell, into which a variety of proteins are inserted. With the lipid phase

DOI: 10.1002/masy.200550104

impermeant to aqueous media, the protein components primarily confer selective and activatable translocation features to the membrane, giving the cells a dynamic capacity to respond appropriately to the environment and signals in it. The majority of membrane research activity has been directed to understanding the operation of the proteins in the membrane, with a significantly smaller effort directed toward understanding the lipid envelop. More recently, however, there has been a shift in interest toward the lipid envelop, driven largely by the recent discoveries of membrane lipid domains, especially rafts and caveolae. Singer and Nicholson (29) first proposed the modern view of the membrane as a phospholipid bilayer as a uniform two dimensional solvent system in which membrane proteins freely floated in a fluid sea of phospholipids. In addition to phospholipids, the membrane also contains substantial amounts of unesterified (free) cholesterol, up to nearly half the total lipid. Rather than being a homogenous mixture of lipids, however, we now know the membrane contains a variety of lipid domains with asymmetric distribution of phospholipids and cholesterol among and between the two membrane leaflets. Currently, its most intriguing features are the various lipid domains which include rafts and caveolae, and several excellent reviews have been published on these structures (20, 21, 28, 30). Less well known are the recently discovered crystalline cholesterol domains which exist in the plane of the membrane and the events leading to their appearance (35). Accordingly, this review focuses on the development of these domains, beginning with phenomenon of biological increases in membrane cholesterol, how this effects membrane structure and cell function, and the obligate phase separation of cholesterol from the phospholipids that leads to the formation of cholesterol domains within the membrane.

The Role of Cholesterol in the Cell Membrane

A. Permeability. In eukaryotes, cholesterol is contained in all cell membranes, but it is present in the highest concentration in the cell's plasma membrane (14). To the average scientist, cholesterol is generally thought to be a benign and inert structural lipid that "sets" membrane fluidity at some level appropriate for the specific membrane and cell type. Indeed there is some truth to this, but setting the fluidity may not be a primary function. Instead, studies have shown that when membranes lack cholesterol they tend to leak considerably, i.e., both water and small molecules (i.e., sizes up to sugars) freely pass. This is thought to result from the transient formation of fractional free volumes across the bilayer leaflets that result from the rotation of the fatty acyl chains of neighboring phospholipids. At various points in time during their rotation, transient gaps or openings appear across the leaflet when the free ends of neighboring acyl chains are at their greatest distance from each other. When this happens simultaneously on both sides of the bilayer, a "fractional free space" forms that briefly allows water and small molecules to penetrate the bilayer (32, 33). The formation of these transient spaces is random and occurs all over the cell membrane. Adding cholesterol to the membrane restricts motion of the fatty acyl chains, condenses the fatty acyl chain region (hydrocarbon core) and greatly decreases membrane permeability. Thus, an essential role for cholesterol in the plasma membrane is to make this membrane particularly impermeant to water and charged particles. This explains why the plasma membrane contains most of the cell's cholesterol load (14). One of the byproducts of this effect of cholesterol is an obligate decrease in membrane fluidity since fluidity is driven by the ability of phospholipid fatty acyl chains to rotate freely.

B. Membrane width. Another byproduct of the presence of cholesterol in the membrane was discovered accidentally in our laboratory in collaborative studies using small angle x-ray scattering (SAXS) analysis (5). The original intent was to determine where in the membrane excess cholesterol resides. In this study a microsomal membrane pellet enriched with plasma membranes was freshly prepared from aortic smooth muscle cells (SMC) isolated from rabbits fed a high cholesterol diet for varying durations. Subjecting these membrane pellets to SAXS analysis revealed an increase in the cholesterol content in the bilayers spanning a dimension extending approximately 9 to 15 Å out from the center of the bilayer (Fig. 1), essentially similar to data obtained from NMR studies (16). Surprisingly, we also found that the phosphorous head-groups spread apart as the cholesterol content increased (Fig. 1 and Fig. 2). This suggested that increasing the cholesterol content of the membrane increased bilayer width. The absolute amount of increase in width is difficult to ascertain since the *d*-space increased approximately 4Å and 5.6 Å while the phosphorus-head-group separation increased 2.5Å, and 4.0Å at 8 and 10 weeks respectively on the cholesterol diet and (Fig. 2). The larger expansion in *d*-space measurement is likely due to hydration differences rather than absolute differences in membrane thickness (22, 34). In our studies, all data were obtained from samples studied at 97% RH and 37° C in order to minimize variations in *d*-space. Phosphorus head-group separation, on the other hand, is less affected by hydration and likely gives a better estimate of the relative differences in membrane thickness observed in membranes isolated at different times on the cholesterol diet. Clearly, the data clearly support the concept that increases in membrane cholesterol content, either in vivo or in vitro, have a membrane "swelling" effect. Note also the tight correlation between membrane cholesterol content and phosphorus head-group spacing as well as *d*-space. Taken together, these data are consistent with a cholesterol-induced increase in bilayer width, as has been empirically demonstrated (18) and implied (23) previously in model systems. This increase in membrane cholesterol content correlated with time on diet, and membrane width correlated with membrane cholesterol content. It is important to note that these data were obtained in membranes that were enriched with cholesterol in vivo, i.e., in rabbits fed a cholesterol-rich (0.5%) diet. This effect of cholesterol on membrane width is therefore a natural biological phenomenon. Moreover, it appears to be a generic effect on phospholipid bilayers since increasing the cholesterol content increases bilayer width in a concentration-dependent fashion in synthetic phosphatidylcholine (PC) bilayers with either a uniform (DMPC) or a biological mixed (BCPC) fatty acyl chain population (Fig. 3).

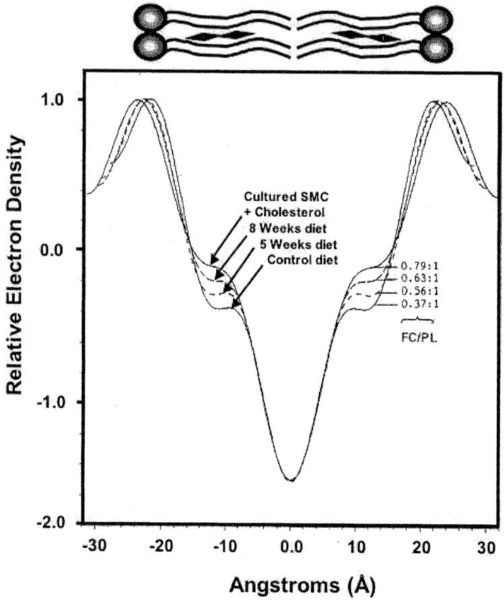

Fig. 1. Electron density profiles of a series of membranes (97% RH, 37°C) isolated from aortic SMC freshly obtained from rabbits maintained on a diet enriched with 0.5% cholesterol for 0, 5 and 8 weeks compared to cultured aortic SMC enriched with exogenous cholesterol. Reproduced from ref.(5).

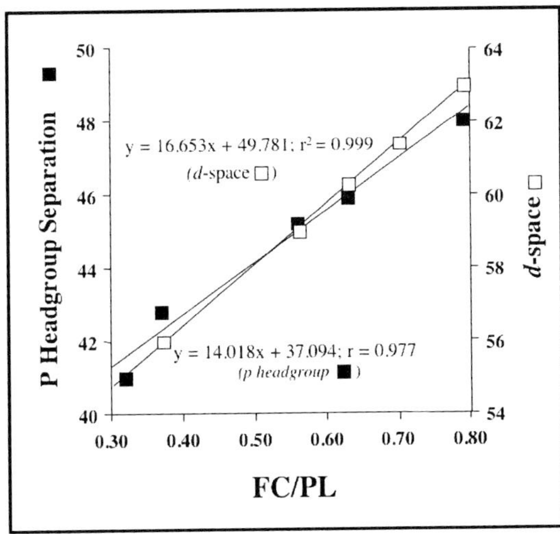

Fig. 2. Membrane bilayer width assessed by *d*-space and phosphorus head-group separation (Å) correlates ($p<.001$) with increasing membrane cholesterol content, and time on the high cholesterol diet. Cells were enriched with cholesterol in vivo or in cell culture (97% RH, 37°C). Redrawn from ref. [Chen, 1995 #13].

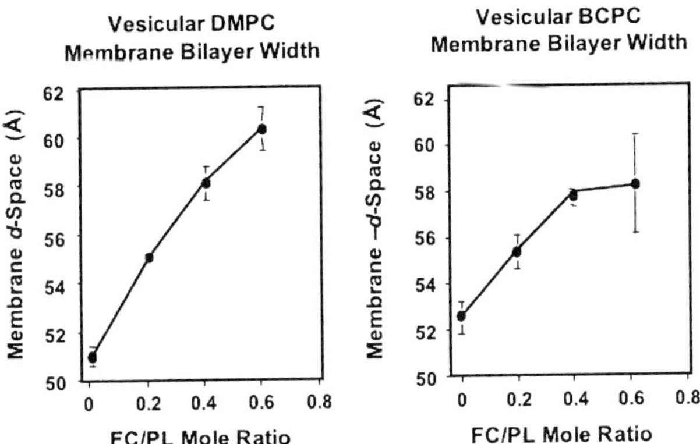

Fig. 3. A. Membrane width increases in a concentration-dependent fashion with increasing cholesterol content in model membranes (35). The left panel illustrates this effect in membranes made with dimyristoylPC (DMPC). The right panel illustrates this effect in membranes made with bovine cardiac PC (BCPC) which contains a biologic mixture of hydrocarbon chains. All samples studied at 97% RH, 37°C. Reproduced from ref. (35) with permission.

Table 1. Phenotypic changes in SMC after cholesterol enrichment *in-vitro* using liposomes (48 hours), *in-vivo* by cholesterol feeding (10 weeks), and the reversal of *in-vivo* effects using human HDL_3.

Phenotypic Alterations in SMC	Induced *in-vitro* by cholesterol	Induced *in-vivo* by cholesterol feeding	Reversal by HDL
1. ⇑ Membrane cholesterol content	√	√	√
2. ⇑ Membrane bilayer width	√	√	√
3. ⇑ Calcium permeability	√	√	√
4. ⇓ ATP-dependent K^+ efflux	√	√	--
5. ⇓ Na^+/K^+ ATPase activity	√	√	√
6. ⇑ Cytosolic calcium levels	√	√	√
7. ⇑ Vasoconstriction	√	√	--
8. ⇑ SMC proliferation	√	√	√
9. ⇑ Mitogen secretion	√	--	√
10. ⇑ Collagen synthesis	√	√	√

Interestingly, the increase in membrane cholesterol content and subsequently membrane width preceded the appearance of atherosclerotic lesions in cholesterol-fed animals by several weeks, suggesting a causal effect of altered membrane structure on the genesis of atherosclerotic vascular disease. We also assessed functions perturbed by cholesterol enrichment (Table 1), focusing on the activity of a variety of membrane proteins including Na^+/K^+ATPase, L-type calcium channels, ATP-dependent K^+ channels. All were disturbed by cholesterol enrichment. Supporting the idea that these alterations were indeed secondary to cholesterol enrichment, all were reversible by reducing membrane cholesterol back to normal using human HDL (HDL_3) as an acceptor (5) The implication that cholesterol directly affects bilayer width is profound. It is well established that the cell's plasma membrane has the greatest cholesterol content (14), as well as the greatest width (2). The width of various cell membranes is estimated accurately from protein hyrophobicity maps, and since the membrane-spanning domains of the various membrane proteins is quite precise in dimensions, it is reasonable to conclude that a particular membrane requires a prespecified width in order to accommodate the specific proteins targeted to that membrane. Accordingly, alterations in membrane cholesterol content, and thus membrane width, would be expected to have major effects on the function of important membrane proteins as we, and others have shown for the membrane-bound Na^+/K^+ATPase (3, 5, 38), the L-type calcium channel (1, 7), the ATP-dependent K^+ efflux channel (24), and the chloride channel (17). In fact, every membrane protein we have examined so far is altered by cholesterol enrichment. Thus, cholesterol appears to have a dual role in the plasma membrane, one of maintaining the bilayer impermeant while simultaneously setting a specific, precise and unique bilayer width.

The Role of Membrane Cholesterol in Cardiovascular Disease

A. Vascular pathology. The above studies clearly demonstrate that excess membrane cholesterol content has major effects on membrane structure and function. Considering that membrane cholesterol content in SMC increases in vivo with dietary hypercholesterolemia (5), as do endothelial cells (15) and cardiac myocytes (9), the question of a pathogenic effect of excess membrane cholesterol becomes compelling. For example, in SMC cholesterol enrichment induces the modulation of these cells from the healthy quiescent phenotype to the atherosclerotic (fibroproliferative) phenotype as reflected by the cholesterol-induced increase in SMC proliferation, collagen synthesis and lesion formation (11). Notably, all these alterations can be reversed by amlodipine, a drug which restores membrane bilayer structure to back to normal and inhibits the development of atherosclerotic lesions (12). In these studies, while amlodipine restored the membrane, it did so without affecting membrane cholesterol content or fluidity, suggesting that the primary alteration in the membrane was structure, not fluidity. Cholesterol enrichment of EC also induces their modulation from the healthy quiescent phenotype to the atherosclerotic (inflammatory and adhesive) phenotype as reflected by the cholesterol-induced activation of NF-kB (15) and/or AP-1 (39) plus monocyte adhesion secondary to ICAM and VCAM expression. To be sure, this view of excess membrane cholesterol as atherogenic fits nicely into the reverse cholesterol transport role for HDL in reducing risk for atherosclerotic syndromes by reducing tissue and cell levels of cholesterol.

Another interesting and novel effect of excess membrane cholesterol potentially contributing to atherogenesis occurs when the cholesterol content of membranes exceeds saturation levels. This occurs in vivo when cholesterol-fed rabbits are maintained on diet beyond 8 weeks. Notably, SAXS analysis reveals phase separation of cholesterol in the membrane as suggested by the appearance of unexpected, highly coherent Bragg's peaks (1' and 2'; Fig 4 top right) (35). Unlike the expected Bragg's peaks with a calculated *d*-space of approximately 56Å (Fig 4 top left), the new peaks had a calculated *d*-space of 34Å. Assuming a cholesterol long axis of 17Å, we concluded that the new Bragg's peaks likely represent diffraction from immiscible tail-to-tail

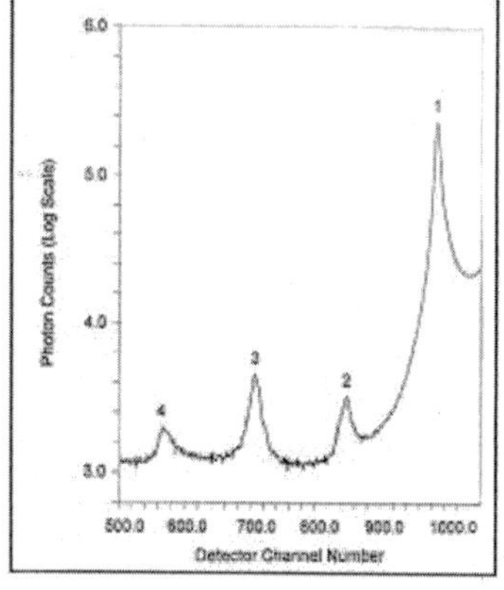

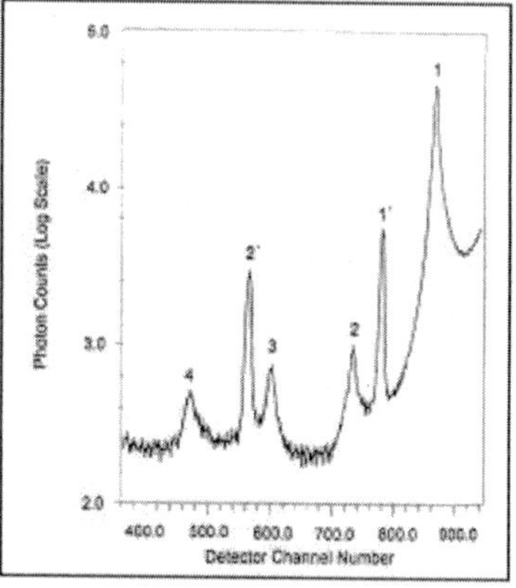

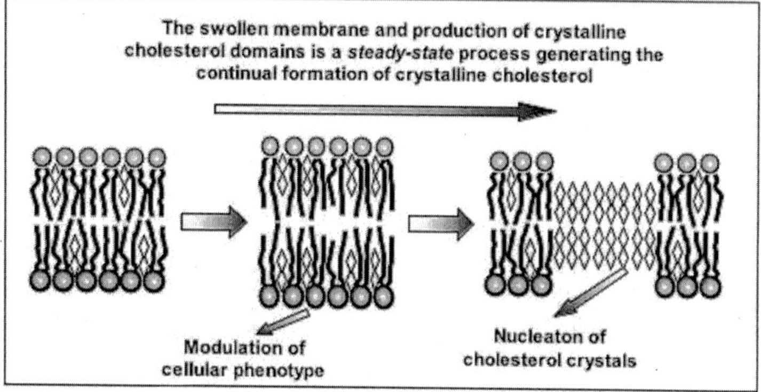

Fig. 4. Top panel shows typical Bragg SAXS peaks obtained from aortic SMC membranes isolated from control animals (left) and animals fed cholesterol for 10 weeks (right) (RH 97%; 37°C) (35). The bottom panel depicts our model of the structural changes in the membrane bilayer with increasing cholesterol concentrations and its consequences. Reproduced from ref. (35) with permission from the publisher.

oriented crystalline cholesterol domains. Support for the presence of these structures within the plane of the membrane (Fig. 4 bottom) comes from studies in which heating from 37° C to 45° C melted these structures with a concomitant increase in bilayer width (35). Likewise, cooling the bilayers back to 37° C resulted in the re-formation of the cholesterol domains as evidence by the reappearance of the 1' and 2' Bragg peaks. Ruocco and Shipely (25) reported similar tail-to-tail oriented crystalline cholesterol domains previously in synthetic membrane bilayers. Our discovery of these structures was the first to document their formation and existence in vivo. Interestingly, Jacobs, et al., (10) later reported similar immiscible cholesterol domains in the lens of the human eye, a tissue whose membranes are highly enriched with cholesterol, further supporting their presence in nature. The finding of immiscible cholesterol domains in membranes of vascular cells was not expected. Their existence, however, may explain a previously unanswered question regarding the origin of crystalline cholesterol "clefts" in atherosclerotic lesions (Fig 5 top) that were described in the medical literature over 70 years ago. Light microscopy typically reveals needles of cholesterol deep inside atherosclerotic plaques in the coronary and peripheral arteries. Since enrichment of the cell membrane with cholesterol is likely a continuous, steady-state condition in hypercholesterolemic subjects, the immiscible cholesterol domains could serve as a site for the continuous nucleation of crystalline cholesterol in atherosclerotic vessels. These crystals have been observed forming inside as well as outside of cells. A similar mechanism has been suggested to be the origin of gall stones (36), structures well known to be rich in unesterified cholesterol. The association of numerous macrophages with the microscopic crystals (Fig. 5B left) suggests that they aggravate lesion formation in affected vessels by attracting inflammatory cells. While the evidence presented above suggests that excess membrane cholesterol may be an important contributor to atherogenesis, is important to point out that this view is clearly forged well "outside the box" of contemporary thinking. While hypercholesterolemia has been long known to induce the development of atherosclerotic lesions in humans and a variety of animal models, the cellular basis for this action virtually ignores the molecule cholesterol. Instead, it attributes atherogenesis to the formation of oxidized LDL (oxLDL) and oxidative injury to the endothelium (31) and SMC. In this context, blood cholesterol therefore is generally viewed as a "marker" for LDL levels and thus oxLDL-mediated oxidative stress. While there appears to be little doubt that oxidative stress plays an important role in atherogenesis, we provide data in this paper consistent with the suggestion that cholesterol enrichment of SMC may contribute importantly to the cellular events that initiate early cellular derangements leading to the development of atherosclerotic lesions. The concept of an atherogenic stimulus secondary to excess membrane cholesterol may identify a redundant pathway to atherogenesis independent of oxidative stress. Accordingly, this may explain the relative ineffectiveness of supplemental dietary antioxidants to provide atheroprotection in human studies (8, 19, 40). On the other hand, the ineffectiveness of anti-oxidant therapy to prevent the development of atherosclerotic lesions in humans may indicate that oxidative stress does not play a central or pivotal role in the pathogenesis of this disease. In this case, a role for excess membrane cholesterol in atherogenesis may be a central issue.

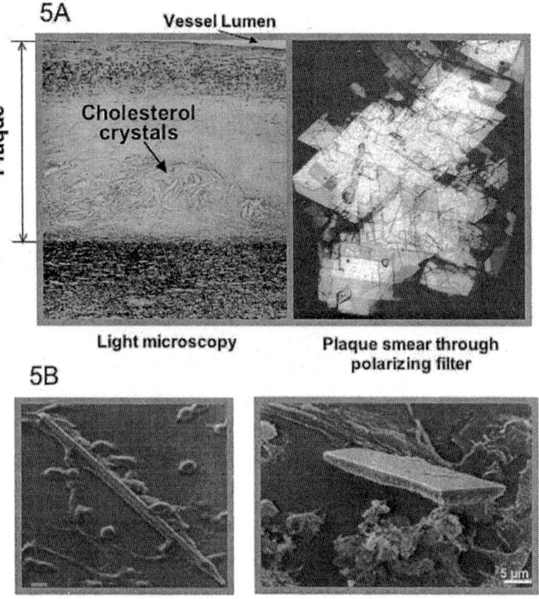

Fig. 5. A. Left panel shows the presence of crystals of cholesterol ("cholesterol clefts") that typically appear in mature atherosclerotic lesions shown under light microscopy of a section of human coronary artery obtained at post-mortem. Right panel shows the crystalline nature of this material when it is spread on a cover slip and viewed under polarizing microscopy. Reproduced from ref. (26) with permission from the publisher. B. Formation of cholesterol crystals in a cultured macrophage cell line enriched with cholesterol. Note the attraction of monocytes to the crystal in the left panel. Reproduced from ref. (12) with permission from the publisher.

B. Cardiac pathology. The heart disease (cardiomyopathy) typically associated with elevated blood cholesterol levels has always been assumed to be strictly limited to coronary artery disease, i.e., an ischemic condition (myocardial ischemia) of the heart muscle (myocardium) which occurs following a sudden occlusion of a coronary artery due rupture of an atherosclerotic plaque. There is no question that people with elevated cholesterol levels are at greatest risk for acute myocardial infarction. We have recently shown in rabbits however, that as blood cholesterol levels rise, membrane cholesterol content in the cardiac myocyte rises simultaneously (Fig. 6A), an effect

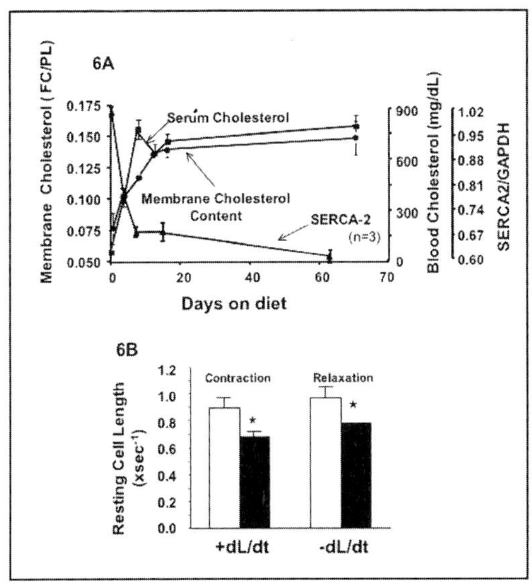

Fig. 6. A. Time course of increase in blood cholesterol in rabbits following the initiation of cholesterol feeding (0.5%). Note the parallel increase in blood and membrane cholesterol content of the cardiac myocyte, and the reciprocal fall in SERCA 2 mRNA gene expression levels.
B. Impaired systolic (contraction) and diastolic (relaxation) function in single left ventricular myocytes freshly isolated from rabbits fed a cholesterol-rich diet (solid bars) compared to control diet (solid bars). Reproduced from ref (9) with permission from the publisher.

that is accompanied by contractile dysfunction (Fig. 6B), i.e., heart failure (9). The contractile impairment is an intrinsic cellular defect which we termed the "cholesterol cardiomyopathy," and is secondary to cholesterol enrichment of the myocyte membrane. This cardiomyopathy is virtually unrelated to any atherosclerotic processes. This was apparent since mRNA expression levels for SERCA 2, a major intracellular calcium regulatory enzyme fell in parallel with the increase in blood and myocyte membrane cholesterol levels, i.e., within only 4 days after the initiation of feeding the cholesterol diet and weeks before arterial disease develops (Fig. 6A). SERCA 2 is a calcium transport enzyme that pumps calcium from the cytosol into the sarcoplasmic reticulum (SR) and thus participates in regulating cardiac contraction and relaxation, both of which were impaired in rabbits fed the high cholesterol diet (Fig 6B). While the magnitude of cardiac dysfunction was mild, it has been shown that hypertension, the leading cause of heart failure in humans, impacts the myocardium to a much greater degree in the presence of hypercholesterolemia than in its absence (37). Considering that over 37 million Americans have hypertension in combination with hypercholesterolemia, this may be an important concern. That the cholesterol cardiomyopathy may exist as a clinical entity in humans is supported by the recent observations that statins, drugs that lower blood cholesterol levels, improve cardiac function in patients with heart failure (6, 13, 27).

Conclusion

No molecule has been more decorated than cholesterol, with no less than 13 Nobel Prizes awarded for studies directed toward understanding its complex chemistry and biology (4). Among its least understood actions are in cell function where it is largely thought of as an inert molecule that provides structural integrity to cell membranes. Studies from our laboratories clearly show that the molecule cholesterol is anything but inert; in fact, it's quite the opposite. Too much cholesterol in the membrane gives rise to a sequence of events that leads to phenotypic modulation to a disease phenotype. In vascular cells, excess membrane cholesterol induces modulation of the cells to the atherosclerotic phenotype, whereas in cardiac myocytes, this leads to a heart failure phenotype. Considering the biological consequences of excess membrane cholesterol, one would anticipate that rigid cellular controls would be in place to firmly hold membrane cholesterol at levels appropriate to preserve the integrity of the plasma membrane. On the contrary, however, membrane cholesterol is rather easily disturbed, both experimentally at the lab bench as well as in vivo in experimental animals as we have repeatedly shown. Moreover, there is considerable evidence that control of membrane cholesterol is equally fragile in humans as well, given the extent of vascular disease in the human population. Figure 4 (bottom) summarizes the potential roles of excess membrane cholesterol to form a "membrane lesion" which initiates and maintains membrane swelling and its progression to the formation of immiscible cholesterol domains. A compelling question now is how excess membrane cholesterol and its effects on membrane structure mediate the rapid changes in gene expression patterns that we have observed. Upstream signaling is likely involved, but how this is accomplished is not clear and constitutes an important area for future studies.

[1] Bialecki RA and Tulenko TN. Excess membrane cholesterol alters calcium channels in arterial smooth muscle. *American Journal of Physiology* 257: C306-C314, 1989.
[2] Bretscher M and Munro S. Cholesterol and the Golgi apparatus. *Science* 261: 1280-1281, 1993.
[3] Broderick R, Bialecki R, and Tulenko TN. Cholesterol-induced changes in rabbit arterial smooth muscle sensitivity to adrenergic stimulation. *American Journal of Physiology* 257: H170-H178, 1989.
[4] Brown MS and Goldstein JL. A receptor-mediated pathway for cholesterol homeostasis. *Science* 232: 34-37, 1986.
[5] Chen M, Mason RP, and Tulenko TN. Atherosclerosis alters composition, structure and function of arterial smooth muscle plasma membranes. *Biochim Biophys Acta* 1272: 101-112, 1995.
[6] de Lorgeril M, Salen P, Bontemps L, Belichard P, Geyssant A, and Itti R. Effects of lipid lowering drugs on left ventricular function and exercise tolerance in dylipidemic coronary patients. *J Cardiovasc Pharmacol* 33: 473-478, 1999.
[7] Gleason MM, Medow MS, and Tulenko TN. Excess membrane cholesterol alters calcium movements, cytosolic calcium levels, and membrane fluidity in arterial smooth muscle cells. *Circ Res* 69: 216-227, 1991.
[8] Henneckens C, Buring J, Manson J, Stampfer M, Rosner B, and et al. Lack of effect of long-term supplementation with beta carotene on the incidence of malignant neoplasms and cardiovascular disease. *N Engl J Med* 334: 1145-1149, 1996.
[9] Huang Y, Walker K, Hanely F, Narula J, Houser S, and Tulenko T. Impaired systolic and diastolic function following a cholesterol-rich diet. *Circulation* 109: 97-102, 2004.
[10] Jacob R, Cenedella R, and Mason R. Evidence for distinct cholesterol domains in fiber cell membranes from cataractous human lenses. *J Biol Chem* 276: 13573-13578, 2001.
[11] Kahn K, Battaglia K, Stepp D, Petrov A, Huang Y, Mason R, and Tulenko T. Influence of serum cholesterol on atherogenesis and intimal hyperplasia after angioplasty: Inhibition by amlodipine. *Am J Physiol (HEART & CIRC)*: (in press), 2004.
[12] Kellner-Weibel G, Yancey PG, Gerome W, Walser GT, Mason RP, Phillips MC, and Rothblat GH. Crystallization of free cholesterol in model macrophage foam cells. *Arterioscler Thromb Vasc Biol* 19: 1891-1898, 1999.
[13] Kjekshus J, Pedersen T, Olsson A, Faegeman O, and Pyorala K. The effects of simvastatin on the incidence of heart failure in patients with coronary artery disease. *J Card Fail* 3: 249-254, 1997.
[14] Lange Y, Swaisgood MH, Ramos BV, and Steck TL. Plasma membranes contain half the phospholipid and 90% of the cholesterol and sphingomyelin in cultured human fibroblast. *J Biol Chem* 264: 3786-3793, 1989.
[15] Laury-Kleintop L, Luo R, Pilane C, and Tulenko T. Increased monocyte adhesion, ICAM-1 and VCAM-1 expression and NF-kB activation in human endothelial cells by cholesterol. *FASEB J* 15: A451, 2001.
[16] Leonard A and Dufourc EJ. Interactions of cholesterol with the membrane lipid matrix. A solid state NMR approach. *Biochimie* 73: 1295-1302, 1991.
[17] Levitan I, Christian A, Rothblat G, and Tulenko T. Alterations in cell volume regulation in endothelial cells following cholesterol enrichment of the cell membrane. *J Gen Physiol* 115: 405-416, 2000.
[18] McIntosh T. The effect of cholesterol on the structure of phosphatidylcholine bilayers. *Biochim Biophys Acta* 513: 43-58, 1978.
[19] Ommen G, Goodman G, Thornquist M, Balmes J, and Culien M. Effects of a combination of β-carotene and vitamin A on lung cancer and cardiovascular disease. *N Engl J Med, :,* 334: 1150-1155, 1996.
[20] Pike L. Lipid rafts: bringing order to chaos. *J Lipid Res* 44: 655–667, 2003.
[21] Pike L. Lipid rafts: heterogeneity on the high seas. *Biochem J* 378: 281–292, 2004.
[22] Rand R and Parsegian V. Hydration forces between phospholipid bilayers. *Biochim Biophys Acta* 988: 351–376, 1989.
[23] Ren J, Lew S, Wang Z, and London E. Transmembrane orientation of hydrophobic alpha-helices is regulated both by the relationship of helix length to bilayer thickness and by the cholesterol concentration. *Biochemistry* 36: 10213-10220, 1997.
[24] Rock D and Tulenko T. Atherosclerosis alters ATP-dependent K^+ channels in arterial smooth muscle. *FASEB Journal* 5: A532, 1991.
[25] Ruocco MJ and Shipley GG. Interaction of cholesterol with galactocerebroside and galactocerebroside-phosphatidylcholine bilayer membranes. *Biophys J* 46: 695-707, 1984.
[26] Sandritter W and Wartman W. *Color Atlas of Histopathology*. Chicago, IL: Year Book Medical Publishers, 1967.
[27] Segal R, Pitt B, Poole-Wilson P, Sharma D, Bradstreet D, and Ikeda L. Effects of HMGCoA reductase inhibitors (statins) in patients with heart failure (abst). *Eur J Heart Failure* 2 Suppl 2: 96, 2000.
[28] Simons K and Toomre D. Lipid rafts and signal transduction. *Nat Rev Mol Cell Biol* 1: 31-41, 2000.
[29] Singer S and Nicholson G. The fluid mosiac model of the structure of cell membranes. *Science* 175: 720-731, 1972.

[30] Smart E, Graf G, McNiven M, Sessa W, Engelman J, Scherer P, Okamoto T, and Lisanti M. Caveolins, liquid-ordered domains, and signal transduction. *Mol Cell Biol* 19: 7289–7304, 1999.
[31] Steinberg D. Low density lipoprotein oxidation and its pathobiological significance. *J Biol Chem* 272: 20963–20966, 1997.
[32] Straume M and Litman B. Equilibrium and dynamic bilayer structural properties of unsaturated acyl chain phosphatidylcholine-cholesterol-rhodopsin recombinant vesicles and rod outer segment disk membranes as determined from higher order analysis of fluorescence anisotropy decay. *Biochemistry* 27: 7723-7733, 1988.
[33] Straume M and Litman B. Influence of cholesterol on equilibrium and dynamic bilayer structure of unsaturated acyl chain phosphatidylcholine vesicles as determined from higher order analysis of fluorescence anisotropy decay. *Biochemistry* 26: 5121-5126, 1987.
[34] Tristram-Nagle S and Nagle J. Lipid bilayers: thermodynamics, structure, fluctuations, and interactions. *Chem Phys Lipids* 127: 3–14, 2004.
[35] Tulenko TN, Chen M, Mason PE, and Mason RP. Physical Effects of cholesterol on arterial smooth muscle membranes: Evidence of immiscible cholesterol domains and alterations in bilayer width during atherogenesis. *J Lipid Res* 39: 947-956, 1998.
[36] Wrenn S, Gudheti M, Veleva A, Kaler E, and Lee S. Characterization of model bile using fluorescence energy transfer from dehydroergosterol to dansylated lecithin. *J Lipid Res* 42: 923-934, 2001.
[37] Wu J-H, Hagaman J, Kim S, Reddick R, and Maeda N. Aortic constriction exacerbates atherosclerosis and induces cardiac dysfunction in mice lacking apolipoprotein E. *Arterioscler Thromb Vasc Biol* 22: 469-475, 2002.
[38] Yeagel PL, Young J, and Rice D. Effects of cholesterol on (Na^+,K^+)-ATPase ATP hydrolyzing activity in bovine kidney. *Biochemstry* 27: 6449-6452, 1988.
[39] Yuan Y, Verna L, Wang N, Liao H, Ma K, Wang Y, Zhu Y, and Stemerman M. Cholesterol enrichment upregulates intercellular adhesion molecule-1 in human vascular endothelial cells. *Biochim Biophys Acta* 1534: 139-148, 2001.
[40] Yusuf S, Dagenais G, Pogue J, Bosch J, and Sleight P. Vitamin E supplementation and cardiovascular events in high-risk patients. The Heart Outcomes Prevention Evaluation (HOPE) Study Investigators. *N Engl J Med* 342: 154-160, 2000.

Macromol. Symp. **2005**, *219*, 39-50

Alteration of Lipid Membrane Rigidity by Cholesterol and Its Metabolic Precursors

Horia I. Petrache, Daniel Harries, V. Adrian Parsegian*

Laboratory of Physical and Structural Biology, National Institute of Child Health and Human Development, National Institutes of Health, Bethesda, MD 20892, USA
E-mail: petrachh@mail.nih.gov

Summary: Caused by biosynthesis defects, cholesterol deficiency can lead to developmental disorders and malformations, with possible implication of lipid membrane properties. We show that modification of sterol chemical structure alters membrane physical properties significantly. By X-ray diffraction and osmotic stress, we measure changes in the bending rigidity of bilayers containing either cholesterol or one of its metabolic precursors. Membrane elasticity differs dramatically between slightly different sterols and varies in the sequence lanosterol < 7-dehydrocholesterol < lathosterol < cholesterol. We interpret the results in terms of sterol location within lipid structures and modification of lateral stress, a structural feature relevant to interactions within biological membranes. We find that cholesterol is most efficient in enhancing membrane rigidity, a possible clue to why depletion or replacement with other sterols can affect cellular structures.

Keywords: bending rigidity; cholesterol biosynthesis; intrinsic curvature; stress profile; X-ray

Lipid bilayers, cholesterol, and cellular function

To a physical chemist, cellular membranes are molecular alloys: their material properties subtly depend on composition, temperature, and other environmental variables. Do all relevant biomembrane functions rely on pure biochemistry, or do they also depend on physical properties of the molecular assembly?

As a rule of thumb, the larger the fraction of a given membrane component, the more likely it is to modulate a material property. This rule holds for cholesterol. Present in plasma membrane of all mammalian cells at mole fractions as high as 30-50% of total lipid, cholesterol likely acts through non-specific physical properties as well as specific interactions.[1,2]

 DOI: 10.1002/masy.200550105

To a cell biologist, cholesterol biosynthesis[3] is a tightly regulated, multistep chemical process via numerous precursors including lanosterol, lathosterol, desmosterol and 7-dehydrocholesterol. As is widely recognized, excess cholesterol leads to atherosclerosis, cardiovascular diseases, and stroke. Deficiency is also dangerous, leading to serious congenital anomalies and mental retardation in newborns.[4,5] Despite numerous studies of cholesterol and its precursors, detailed understanding of sterol-lipid interactions in relation with membrane architecture is still lacking.

Structural differences between cholesterol and its precursors include the number and position of double bonds and additional methyl groups.

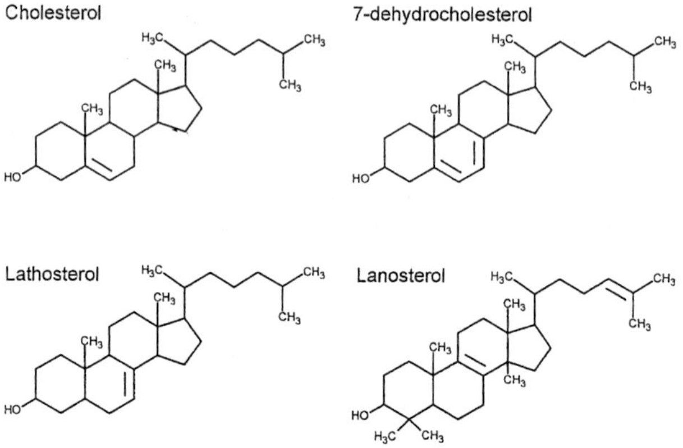

Can bilayer material properties conferred by cholesterol differ significantly from its precursors? To what extent can additional double bonds and methyl groups modify sterol-lipid interactions? To address these questions, we analyze interbilayer interactions and curvatures of hexagonal phases. We follow a simple procedure to obtain indirect information on bending rigidities, and find that cholesterol makes significantly more rigid bilayers than other sterols in its biosynthesis pathway. These results can be of significance to understanding the molecular mechanisms responsible for manifestations of cholesterol-related disorders.

The balance of forces within lipid bilayers

From the heterogenous molecular structure of lipids themselves and from the tumultuous water/lipid headgroup region[6], aggregated lipids are subjected to a non-uniform distribution of forces (lateral stress).[7-10] As depicted in Figure 1, this inhomogenous distribution of forces along the bilayer normal generates bending moments.[11] Were they not paired into bilayers, individual monolayers would generally tend to deviate from planar geometry. A lipid bilayer composed of frustrated monolayers lives under stress. It turns out that the balance of forces within bilayers can easily be offset by additives such as sterols.[1,7]

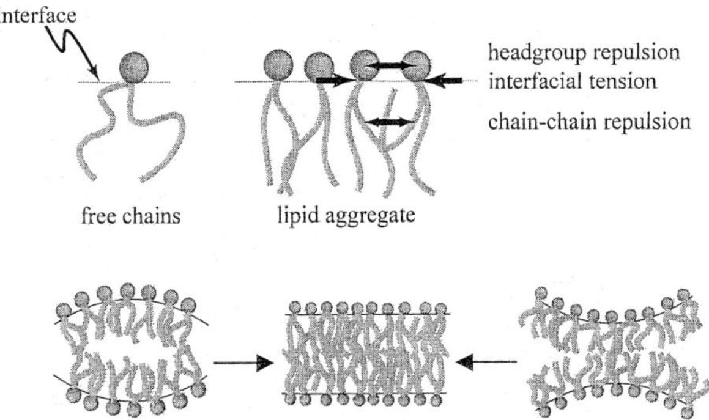

Figure 1. Schematics of the inhomogeneous distribution of lateral forces within lipid bilayers. Monolayer bending tendency (moments) lead to packing frustration.

Within this framework, we can think of sterol effects as changes in monolayer bending moments. This change is measured by the monolayer bending rigidity, K_C^{mono}, and the monolayer intrinsic curvature, $1/R_0$. The intrinsic curvature is the value at which the monolayer free energy is minimum in the absence of other terms, e.g. monolayer coupling. In a harmonic approximation, the free energy per area needed to bend a monolayer away from R_0 is[12],

$$F_{elastic}^{m}(R) \approx \frac{1}{2} K_C^{mono} \left(\frac{1}{R} - \frac{1}{R_0} \right)^2 . \qquad (1)$$

Roughly speaking, K_C^{mono} represents the energy needed to flatten a spontaneously curved monolayer of area $2R_0^2$. To calculate the free energy of bending a *bilayer*, we sum up the bending energies of paired monolayers, taking into account their opposite direction of bending (competing springs). In the thin film approximation, the bilayer bending rigity has the form

$$K_C \cong 2K_C^{mono}[1+\vartheta(D_B/R_0)], \qquad (2)$$

where $\vartheta(D_B/R_0)$ indicates a correction on the order of D_B/R_0. The bilayer bending rigidity is larger than twice the rigidity of a monolayer, a stiffening effect due to the finite thickness D_B. For many bilayer-forming phospholipids, K_C is on the order of $\sim 10^{-12}$ erg $\approx 23\ k_BT$[13].

The method of choice here to probe this property is the measurement of the response of lipid aggregates to osmotic stress.[14-16] Multilamellar lipid bilayers are hydrated in the presence of osmolytes, e.g. high-molecular weight polyethyleneglycol (PEG), and interbilayer separation is measured, using X-rays, as a function of the applied osmotic stress set by the PEG/water weight ratio. These measurements produce force-distance curves analogous to pressure-volume curves of fluids, as will be shown below.

X-ray images of lipid phases

To study sterol effects on bending moments and intrinsic curvatures, we use lipids which, depending on their intrinsic curvature values, self-assemble into either lamellar (L_α) or inverted hexagonal (H_{II}) phases at physiological temperatures. Lipids with saturated acyl chains and phosphatidylcholine (PC) headgroups form bilayer phases. Conversely, lipids with unsaturated chains and phosphatidylethanolamine headgroups (PE, a demethylated PC) form inverse hexagonal phases. We have chosen the 14-carbon disaturated dimyristoylphosphatidylcholine (DMPC) and the monounsaturated 18-carbon dioleoylphosphatidylethanolamine (DOPE) as representative of each class. The geometry of lipid aggregates, and corresponding X-ray pictures are shown in Figure 2. Rings rather than points of scattering are obtained due to random, "powder" orientations of lipid suspensions in the X-ray beam. The lattice (repeat) spacings are determined from the position of these rings. For lamellar structures, the rings are equally spaced and index simply as 1,1/2, 1/3, …, while for the H_{II} phase the indexing is $1, 1/\sqrt{3}, 1/2, 1/\sqrt{7}, \ldots$.[8,12]

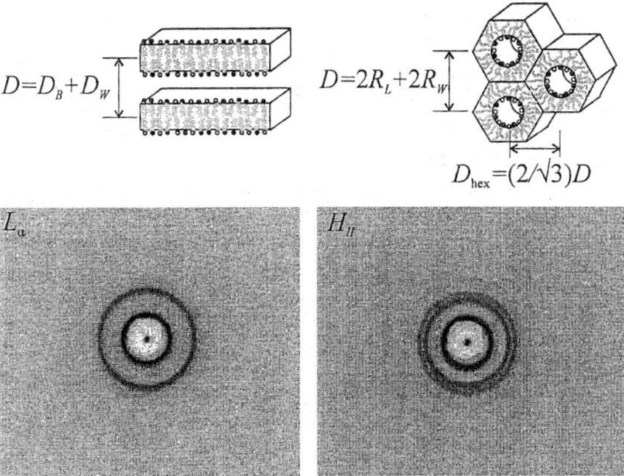

Figure 2. X-ray scattering from multilamellar (L_α) DMPC and inverted hexagonal (H_{II}) DOPE, both fully hydrated at 35°C. Scattering rings of DMPC are equally spaced and reflect the regular spacing between stacked bilayers. The X-ray rings of DOPE index as 1,1/√3, 1/2, 1/√7,... and reflect the honeycomb-like positioning of water cylinders in the H_{II} phase.

DOPE has a negative spontaneous curvature and forms an inverted hexagonal phase when fully hydrated. DMPC, with a much smaller spontaneous curvature, forms a lamellar phase. The structures depicted in Figure 2 typically extend to 10^2-10^3 bilayers (DMPC) or cylinders (DOPE) in a given scattering domain, as determined from the sharpness of the scattering peaks. For fully hydrated DMPC in water at 35°C, the interlamellar repeat spacing is D = 63 Å, which we decompose into a bilayer thickness D_B = 44 Å [16], and a water spacing of D_W = 19 Å (see drawing in Figure 2). For DOPE, with a hexagonal lattice spacing of D_{hex} = 64 Å, D = 74 Å, with $2R_L$ = 36 Å and $2R_W$ = 38 Å[17]. The water content, measured by D_W and R_W, is reduced under osmotic stress.

The highly curved hexagonal phases and the way in which they respond to osmotic stress (dehydration) have been directly related to bending energies.[12,17-19] For the most biologically relevant lamellar geometry, however, in which curvature is not explicit, we obtain information on bilayer elasticity from measurements of membrane shape fluctuations.[16,20-23]

Bilayer interactions

For neutral fluid membranes, the interbilayer spacing is set by the balance of van der

Waals attraction of hydrocarbon slabs in water, and two repulsion forces, termed "hydration" and "shape-fluctuation" forces, respectively.[20,24] The attractive van der Waals force can be calculated analytically for a pair of infinitely extended slabs.[25,26] The repulsive forces are described phenomenologically through empirically determined parameters and functional forms,[16,20,24]

$$F(D_W,T) = -\frac{H}{12\pi D_W^2} + P_h\lambda e^{-D_W/\lambda} + \left(\frac{k_B T}{2\pi}\right)^2 \frac{1}{K_C\sigma^2} \tag{3}$$

$$P_{osm} = -\frac{dF(D_W,T)}{dD_W} \tag{4}$$

Equation 3, an idealized form of separable terms, gives the interaction energy per membrane unit area $F(D_W = D - D_B, T)$, as a function of interlamellar water spacing, D_W, and temperature, T. Equation 4 relates the free energy, F to the applied osmotic pressure, P_{osm}. Experimental results for P_{osm} *vs.* D for DMPC in the fluid state are shown in Figure 3, together with the decomposition into the various interaction terms, adapted from reference 16.

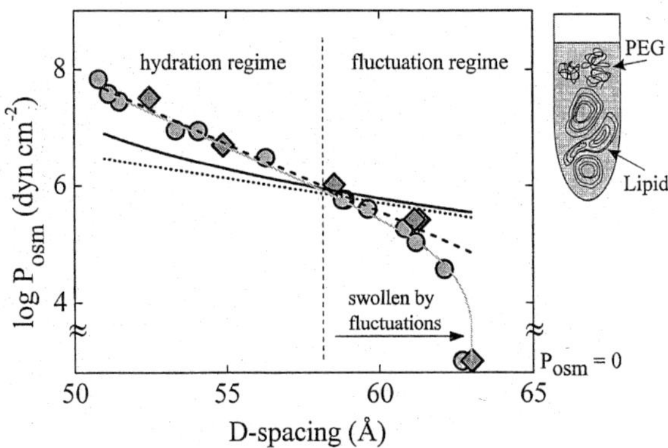

Figure 3. Variation of interlamellar repeat spacing of DMPC multilayers with applied osmotic stress. Diamonds show current data at 35°C and circles data at 30°C from Ref. 16. A hydration and a fluctuation regime are distinguished by the decomposition of forces (Equations 3-4) into van der Waals attraction (solid line), hydration repulsion (dashed), and fluctuation repulsion (dotted).

The first term in Equation 3 represents the attractive van der Waals term (solid line in Figure 3) decaying quadratically with interbilayer separation. The strength of this interaction is given by the Hamaker parameter, H, on the order of $4\text{-}5 \cdot 10^{-14}$ erg ($\approx 1\ k_B T$) for hydrocarbon/lipids in water.[25]

The second term in Equation 3, representing the hydration force (dashed line, Figure 3), accounts for the energetic cost of water ordering in the vicinity of lipid headgroups.[27] This force is exponential, with a decay length of about 2 Å. It therefore weakens significantly as interbilayer spacings increase by 10 Å or more.

The third term in Equation 3, the shape-fluctuation or undulation term (dotted line, Figure 3) acts at larger inter-membrane separation, accounting for the entropic penalty due to confinement of undulating membranes.[28] This entropic force is inversely proportional to the bilayer bending rigidity, K_C. The value used for the plot in Figure 1 is $K_C = 0.8 \cdot 10^{-12}$ erg ($\approx 19\ k_B T$).[16] The parameter σ in Equation 3 is a function of D_W and represents the root mean-square fluctuation in interbilayer separation, $\sigma^2 = \langle D_W^2 \rangle - \langle D_W \rangle^2$, where the brackets indicate the ensemble average over all fluctuation modes. The variation of σ with D_W required by Equation 4 can be measured experimentally by high resolution X-ray diffraction.[16,22] Note that for evaluation of forces, the natural parameter for interbilayer separation is the water spacing $D_W = D - D_B$, rather than the repeat spacing. For simplicity, the qualitative analysis presented here uses the directly measured D.

Figure 3 shows two distinct regimes. A hydration regime exists at high osmotic pressures, where the van der Waals attraction is balanced mainly by the hydration repulsion. The fluctuation force here is negligible. A second region, below 1 atm of osmotic stress, has the fluctuation force as the dominant repulsion term (fluctuation regime); here, the hydration force can be neglected. Van der Waals and hydration forces come into balance at a repeat spacing of about 58 Å (D_W = 14 Å), compared to the full free energy minimum at 63 Å (D_W= 19 Å). Fluctuations enhance swelling by 5 Å (25%).

Applying osmotic stress to sterol-containing bilayers

We can now investigate the modification of bilayer interactions by addition of sterols. In Figure 4 we plot the repeat spacings of DMPC/sterol mixtures versus sterol content. Panel A shows the full swelling at zero osmotic pressure and panel B the reduction under 0.26

atm of osmotic stress (5% PEG solutions, log P_{osm} = 5.4). There are marked differences between the effects of sterols at full hydration (Figure 4A). For all bilayer compositions, the lamellar repeat spacing increases in the order cholesterol < lathosterol < 7-dehydrocholesterol < lanosterol. When fluctuations are suppressed by mild osmotic stress (0.26 atm), these differences are eliminated (Figure 4B).

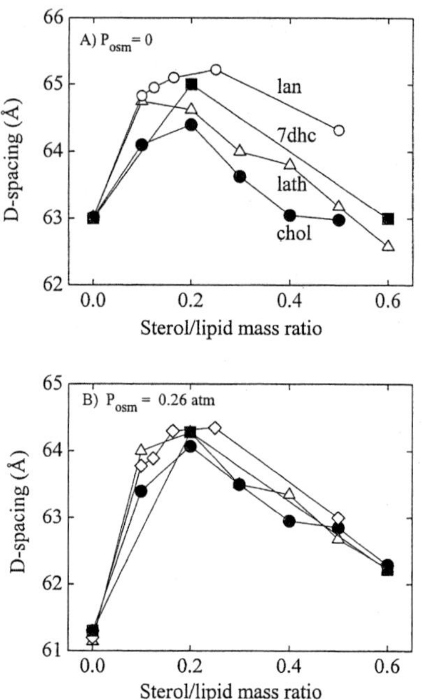

Figure 4. Interlamellar repeat spacing versus sterol content for DMPC multilayers at 35°C. (A) Large differences between sterols are seen for freely fluctuating bilayers at full hydration. (B) Differences vanish when fluctuations are suppressed by mild osmotic stress.

The modifications of interlamellar spacings by low osmotic pressure strongly suggest that sterols have noticeable effects on the bending rigidity of DMPC bilayers. In particular, cholesterol forms the most rigid bilayers (least swelling due to fluctuations), while the trimethylated precursor lanosterol makes most flexible bilayers of all sterols shown.[29-33]

The variation of *D* with sterol content in Figure 4 reflects the phase diagram of DMPC/sterol mixtures.[34-41] At 35°C, a coexistence region exists between a liquid disordered phase (with low sterol content) and a liquid ordered phase (with high sterol content). The sudden decline in the D-spacing values starting at a sterol/lipid mass ratio of 0.2 (30% mole sterol) corresponds to the transition into the liquid ordered phase.[34-41]

Discussion

Motivated by studies of cholesterol disorders, we have investigated the possibility that replacement of cholesterol with other sterols affects membrane bending elasticity, despite similarities in chemical structure. Indeed, seen through differences in forces of bilayer undulations, cholesterol makes significantly more rigid bilayers than its precursors, suggesting a critical role for cholesterol in modulating cellular structure and function.[42,43]

Bilayer fluctuations are suppressed by osmotic stress. For pure DMPC (Figure 3), interbilayer separation is reduced from 19 Å to about 14 Å when fluctuations are completely removed. A simple calculation using Equation 3, indicates that a similar reduction in spacing can be obtained through a 10-fold increase of the bending rigidity, K_C, when all other interaction parameters are held fixed. Measurements of K_C by various experimental methods: micropipette pressurization,[44] fluctuation microscopy,[45,46] high-resolution X-ray diffraction,[16,23] showed less than a 5-fold increase by addition of 50% cholesterol.[1,44-46] This means that bilayer fluctuations are not completely suppressed by sterols, an interpretation supported by the results in Figure 3.

We find that membrane rigidity is significantly reduced by substitution of cholesterol with other sterols, in agreement with fluctuation microscopy[46] and ^{2}H NMR[30,31] measurements. The variations with sterol type in Figure 3 can be explained by a decrease in K_C of about 15 k_BT between cholesterol and the less rigid lanosterol-containing bilayers. This difference is significant. For illustration, it would take 15 k_BT more bending energy to encapsulate a typical size protein in a liposome containing cholesterol than one with lanosterol. This energy could be accounted for by other lipid-protein interactions, for example, an additional 5 net electrostatic charges per liposome provide enough favorable binding energy for encapsulation of an oppositely charged protein.[47] This would amount to an additional 5-10% of charged lipids; in comparison, cell membranes contain about 10-20% charged lipids.

Differences between the sterol effects on membrane rigidity can be rationalized in terms of sterol location within the lipid bilayer. Cholesterol has a significant ordering effect on the lipid acyl chains, as shown by ^{2}H NMR[1,30,40,48] and X-ray diffraction.[49-54] By contrast, only a slight effect has been detected on lipid headgroups,[55] although this may depend on lipid type and temperature[29,56]. Addition of cholesterol could create more space between lipid headgroups, thereby allowing greater headgroup disorder. It is therefore expected that sterols introduce an inhomogenous modification of lateral forces within bilayers (cf. Figure 1). The lateral stress profile and, in particular, the bending moments will therefore depend on the sterol location inside the membrane. For example, more polar sterols (due to additional double bonds) might protrude further into the headgroup/water interface than cholesterol, and render the bilayer more flexible.

We have reached a similar conclusion from measurements of DOPE/sterol mixtures forming H_{II} phases (data not shown). Cholesterol-containing DOPE forms hexagonal structures with a smaller radius of curvature than with other sterols. Quantifying changes in bending rigidity and intrinsic curvature values is in progress. Here we note that according to Equation 2, the membrane bending rigidity K_C can be modified through the monolayer bending rigidity K_C^{mono} as well as the intrinsic curvature value R_0. It is of interest to determine whether sterols affect one or both of these elastic parameters.

The behavior of interlamellar spacing with increased sterol content is not resolved by these data. From the X-ray[50,51] and ^{2}H NMR[1,31,40,48] measurements mentioned above, the bilayer thickness is estimated to increase with sterol by up to 3 Å. However, if bilayer rigidity also increases with added sterol, then a decrease in the interlamellar water spacing is expected due to weaker fluctuation repulsion. The competition between the two effects, added thickness *vs.* fluctuation suppression, could explain the peak of D *vs.* composition in part A of Figure 4. However, because the peak is still present even under osmotic stress in part B of Figure 4, it most likely reflects the variation of the bilayer thickness with increasing sterol content. It is conceivable that at 30% mole fraction (corresponding to 0.2 mass ratio) where the transition into the liquid ordered phase occurs, there is abrupt bilayer reorganization. In the liquid ordered phase, bilayer thickness might decrease with increasing sterol content, even though lipid acyl chains are stretched (as measured by ^{2}H NMR). This can be due to interpenetration of lipid chains from the two monolayers at the bilayer center to accommodate the length mismatch between the cholesterol molecule and the acyl chain of the lipids.[37,51-54,57]

The hydration repulsion can also change with the addition of sterol. This change is suggested by a picture of sterols acting as lateral spacers within the lipid matrix. A weaker hydration repulsion in the liquid ordered phase could also explain the decrease of D at high sterol content in Figure 4, as an alternative to a decrease in bilayer thickness discussed above. Of main interest here, however, is that modification of either hydration forces, or of bilayer thicknesses are independent of sterol type, as shown by the measurements under osmotic stress (Figure 4B). Sterol type dependence appears only in the absence of stress (Figure 4A) suggesting a stronger modification of fluctuations rather than of hydration repulsion.

As seen through modification of fluctuation forces, sterols modify bilayer bending rigidities. It is remarkable that the evolutionary process has led to cholesterol, the sterol forming the most rigid bilayers. It is therefore conceivable that, in addition to specific biochemical mechanisms, cellular manifestations of cholesterol disorders could involve modifications of membrane physical properties.

Acknowledgments

We thank our collaborators, Y. Peng Loh and Marjorie Gondré-Lewis for inspiring discussions on cholesterol disorders, and Emily M. Dykstra, Gary Martinez, and Michael F. Brown for illuminating discussions on NMR measurements and for kindly providing lanosterol samples.

[1] M. Bloom, E. Evans, O. G. Mouritsen, *Q. Rev. Biopys.* **1991**, 24, 293-397.
[2] P. L. Yeagle, Biology of Cholesterol, **1988**, CRC Press, Inc., Boca Raton, FL.
[3] A. L. Lehninger, Biochemistry, 2nd edn., **1975**, Worth Publishers Inc., New York.
[4] J. P. Incardona, S. Eaton, *Curr. Opin. Cell Biol.* **2000**, 12, 193-203.
[5] R. I. Kelley, G. E. Herman, *Annu. Rev. Genomics Hum. Genet.* **2001**, 2, 299-341.
[6] S. H. White, A. S. Ladokhin, S. Jayasinghe, K. Hristova, *J. Biol. Chem.* **2001**, 276, 32395-32398.
[7] S. M. Gruner, *Proc. Natl. Acad. Sci.*, **1985**, 82, 3665-3669.
[8] J. M. Seddon, *Biochim. Biophys. Acta*, **1990**, 1031, 1-69.
[9] D. Harries, A. Ben-Shaul, *J. Chem. Phys.* **1997,** 106 , 1609-1619.
[10] S. M. Bezrukov, *Curr. Op. Coll. Int. Sci.* **2000**, 5, 237-243.
[11] I. Szleifer, D. Kramer, A. Ben-Shaul, W. M. Gelbart, S. A. Safran, *J. Chem. Phys.*, **1990**, 92, 6800-6817.
[12] S. M. Gruner, V. A. Parsegian, R. P. Rand, *Faraday Discuss. Chem. Soc.*, **1986**, 81

[13] W. Rawicz, K. C. Olbrich, T. J. McIntosh, D. Needham, E. Evans, *Biophys. J.*, **2000**, 79, 328-339.
[14] V. A. Parsegian, N. Fuller, R. P. Rand, *Proc. Natl. Acad. Sci.*, **1979**, 76, 2750-2754.
[15] T. J. McIntosh, S. A. Simon, *Biochemistry*, **1986**, 25, 4058-4066.
[16] H. I. Petrache, N. Gouliaev, S. Tristram-Nagle, R. T. Zhang, R. M. Suter, J. F. Nagle, *Phys. Rev. E.* **1998**, 57, 7014-7024.
[17] P. E. Harper, D. A. Mannock, R. N. A. H. Lewis, R. N. McElhaney, S. M. Gruner, *Biophys. J.* **2001**, 81, 2693-2706.
[18] P. Rand, N. L. Fuller, S. M. Gruner, V. A. Parsegian, *Biochemistry*, **1990**, 29, 76-87.
[19] Z. Chen, R. P. Rand, *Biophy. J.* **1997**, 73, 267-276.
[20] E. A. Evans, V. A. Parsegian. *Proc. Natl. Acad. Sci.* **1986**, 83, 7132-7136.
[21] E. B. Sirota, G. S. Smith, C. R. Safinya, R. J. Plano, N. A. Clark, *Science*, **1988**, 242, 1406-1409.
[22] H. I. Petrache, S. Tristram-Nagle, K. Gawrisch, D. Harries, V. A. Parsegian, J. F. Nagle, *Biophys. J.* **2004**, 86, 1574-1586.
[23] S. Tristram-Nagle, J. F. Nagle, *Chem. Phys. Lipids*, **2004**, 3-14.
[24] R. Podgornik, V. A. Parsegian, *Langmuir*, **1992**, 8, 557-562.
[25] V. A. Parsegian, G. H. Weiss, *J. Coll. Int. Sci.*, **1981**, 81, 285-289.
[26] V. A. Parsegian, *Langmuir*, **1993**, 3625-3628.
[27] S. Marčelja, N. Radic, *Chem. Phys. Lett.* **1976**, 42, 129-130.
[28] W. Helfrich, *Z. Naturforsch* **1978**, 33a, 305-315.
[29] J. A. Urbina, S. Pekeerar, H-B. Le, J. Patterson, B. Montanez, E. Oldfield, *Biochim. Biophys. Acta* **1995**, 1238, 163-176.
[30] L. Miao, M. Nielsen, J. Thewalt, J. H. Ipsen, M. Bloom, M. J. Zuckermann, O. G. Mouritsen, *Biophys. J.*, **2002**, 82, 1429-1444.
[31] G. V. Martinez, E. M. Dykstra, S. Lope-Piedrafita, M. F. Brown, *Langmuir* **2004**, 20, 1043-1046.
[32] E. Endress, S. Bayerl, K. Prechtel, C. Maier, R. Merkel, T. M. Bayerl, *Langmuir*, **2002**, 18, 3293-3299.
[33] E. Endress, H. Heller, H. Casalta, M. F. Brown, T. M. Bayerl, *Biochemistry* **2002**, 41, 13078-13086.
[34] P. F. F. Almeida, W. L. C. Vaz, T. E. Thompson, *Biochemistry* **1992**, 31, 6739-6747.
[35] F. Richter, G. Rapp, L. Finegold, *Phys. Rev. E.*, **2001**, 63, 051914.
[36] W. Knoll, G. Schmidt, K. Ibel, E. Sackmann, *Biochemistry*, **1985**, 24, 5240-5246.
[37] T. P. W. McMullen, R. N. A. H. Lewis, R. N. McElhaney, *Biochemistry* **1993**, 32, 516-522.
[38] J. C. Owicki, H. M. McConnel, *Biophys. J.* **1980**, 30, 383-397.
[39] D. Recktenwald, H. M. McConnel, *Biochemistry* **1981**, 20, 4505-4510.
[40] J. L. Thewalt, C. E. Hanert, F. M. Linseisen, A. J. Farrall, M. Bloom, *Acta Pharm.* **1992**, 42, 9-23.
[41] J. H. Ipsen, O. G. Mouritsen, M. J. Zuckermann, *Biophys. J.* **1989**, 56, 661-667
[42] Y. Wang, C. Thiele, W. B. Huttner, *Traffic* **2000**, 1, 952-962.
[43] S. Dhanvatari, Y. P. Loh, *J. Biol. Chem.* **2000**, 275, 29887-29893.
[44] E. Evans, D. Needham, Faraday Discuss. **1986,** 81:267-280.
[45] P. Méléard, C. Gerbeaud, T. Pott, L. Fernandez-Puente, I. Bivas, M. D. Mitov, J. Dufourcq, and P. Bothorel, *Biophys. J.* **1997**, 72, 2616-2629.
[46] J. Henriksen, A. C. Rowat, J. H. Ipsen, *E. Biophys. J.* (in press).
[47] D. Harries. A. Ben-Shaul, I. Szleifer, *J. Phys. Chem.* **2004**, 1491-1496.
[48] M. Lafleur, R. P. Cullis, M. Bloom, *Eur. Biophys. J.* **1990**, 19, 55-62.
[49] P. Rand, V. Luzzati, *Biophys. J.* **1968**, 8, 125-137.
[50] T. J. McIntosh, *Biochim. Biophys. Acta* **1978**, 513, 43-58.
[51] D. Needham, T. J. McIntosh, E. Evans, *Biochemistry* **1988**, 27, 4668-4673.
[52] S. A. Simon, T. J. McIntosh, A. D. Magid, D. Needham, *Biophys. J.* **1992**, 61, 786-799.
[53] J. B. Finean, *Chem. Phys. Lipids* **1989**, 49, 265-269.
[54] J. B. Finean, *Chem. Phys. Lipids*, **1990**, 54, 147-156.
[55] M. F. Brown, J. Seelig, *Biochemistry* **1978**, 17, 381-384.
[56] R. M. Epand, D. W. Hughes, B. G. Sayer, N. Borochov, D. Bach, E. Wachtel, *Biochim. Biophys. Acta* **2003**, 1616, 196-208.
[57] S. May, A. Ben-Shaul, *J. Chem. Phys.* **1995**, 103, 3839-3848.

Modeling the Electrostatics and Size Effect within a Crowded Bioenvironment

Zhidong Li, Jianzhong Wu*

Department of Chemical and Environmental Engineering, University of California, Riverside, CA 92521-0425, USA

Summary: Biological fluids typically contain a large number of macromolecules occupying up to 40% of the total volume. Current understanding of the effect of high concentration, or 'macromolecular crowding', on cellular processes is primarily based on the excluded-volume considerations in which all intermolecular interactions beyond the short-ranged repulsion are neglected. In this work, a density functional theory (DFT) accompanied by Monte Carlo simulations is employed to investigate the structural and thermodynamic properties of a crowded cellular environment within the primitive model where biomacromolecules are represented by neutral and charged hard spheres and the solvent by a continuous dielectric medium. The performance of the DFT has been tested with extensive results from Monte Carlo (MC) simulations for the pair correlation functions (PCFs), excess internal energies, and osmotic coefficients under a variety of solution conditions.

Keywords: density functional theory; macromolecular crowding; Monte Carlo simulation; primitive model

Introduction

Dispersions of macroions and neutral species at high volume occupation provide a simple model to represent the crowded cellular environment that contains numerous biomacromolecules and cellular polymers. The so-called 'macromolecular crowding' affects the functions and activities of biomacromolecules and is relevant to most biological processes in every body cell, in the blood, and in all body fluids, especially the intercellular fluids[1,2]. Without exaggeration, essentially all life processes take place in a crowded environment.

The structural and thermodynamic properties of macromolecular crowding are primarily determined by the excluded-volume effects and the long-ranged Coulomb interactions. A number of theoretical and simulation techniques have been applied to investigating the properties of macromolecular crowding in the context of the primitive model where biomacromolecules are treated as charged and neutral hard spheres and the solvent as a continuous dielectric medium[3]. These techniques range from the conventional Poisson-Boltzmann (PB) equation and the

 DOI: 10.1002/masy.200550106

hypernetted chain (HNC) approximation, to Monte Carlo (MC) simulations. At least within the primitive model, the basic physics is now well understood using conventional approaches. However, quantitative representation of the structural and thermodynamic properties of such systems remains a grand challenge. The difficulty is mainly due to the high asymmetry of the constituents in terms of both size and charge valence. The purpose of present work is to apply a newly developed density functional theory (DFT)[4], along with Monte Carlo simulations, to investigating the structural and thermodynamic properties of macromolecular crowding. Unlike many previous publications for macromolecular crowding, this work accounts for both excluded volume and electrostatic interactions.

Theoretical Background

Within the primitive model for 'macromolecular crowding', the intrinsic Helmholtz free energy functional $F[\{\rho_i(\mathbf{r})\}]$ can be decomposed into four distinctive contributions including an ideal-gas term, the excluded-volume effect, the direct Coulomb interactions, and the intermolecular correlations:

$$F[\{\rho_i(\mathbf{r})\}]=F^{id}[\{\rho_i(\mathbf{r})\}]+F_{hs}^{ex}[\{\rho_i(\mathbf{r})\}]+F_C^{ex}[\{\rho_i(\mathbf{r})\}]+F_{el}^{ex}[\{\rho_i(\mathbf{r})\}]. \quad (1)$$

In Eq.(1), $\{\rho_i(\mathbf{r})\}$ stands for a set of density profiles for all species, i.e., macroions, neutral species and small ions. While the ideal-gas contribution $F^{id}[\{\rho_i(\mathbf{r})\}]$ and the direct Coulomb energy $F_C^{ex}[\{\rho_i(\mathbf{r})\}]$ are known exactly, the other two terms, $F_{hs}^{ex}[\{\rho_i(\mathbf{r})\}]$ and $F_{el}^{ex}[\{\rho_i(\mathbf{r})\}]$, can only be evaluated with some approximations. The excluded-volume term arises from the sizes of small ions and biomacromolecules; it is represented by a modified fundamental measure theory (MFMT). $F_{el}^{ex}[\{\rho_i(\mathbf{r})\}]$ takes into account the intermolecular correlations due to the Coulomb and hard-sphere interactions. This term is calculated by using a quadratic functional Taylor expansion. If $F_{el}^{ex}[\{\rho_i(\mathbf{r})\}]$ vanishes, Eq.(1) leads to the conventional Boltzmann equation for the electrostatic interactions. To evaluate the Helmholtz energy functional due to the intermolecular correlations, we use the direct correlation functions (DCFs) obtained from the mean-spherical approximation (MSA). At equilibrium, the pair correlation functions (PCFs) can be calculated from the Euler-Lagrange equation, and subsequently all thermodynamic properties can be calculated via standard statistical-mechanics relations.

To test the reliability of the analytical results, we have compared the pair distribution functions and internal energies with the corresponding results from the canonical ensemble Monte Carlo simulations. The details of theory and simulations have been given elsewhere[4].

Results and Discussion

We first examine the performance of the DFT for predicting the microscopic structures (as represented by various PCFs) of crowded systems containing macroions, neutral particles and small ions. Throughout this work, the symbols g_{++}, g_{--}, g_{+-}, g_{0+}, g_{0-} and g_{00} designate, respectively, the macroion-macroion, counterion-counterion, macroion-counterion, neutral-macroion, neutral-counterion and neutral-neutral pair correlation functions (PCFs). For all the systems considered in this work, the size ratio of the macroions, counterions and neutral particles is fixed at 4.0nm/0.4nm/1.6nm, and the concentration of electrolyte C_e and the valence of macroions Z_+ are retained at 0.002M and +15, respectively.

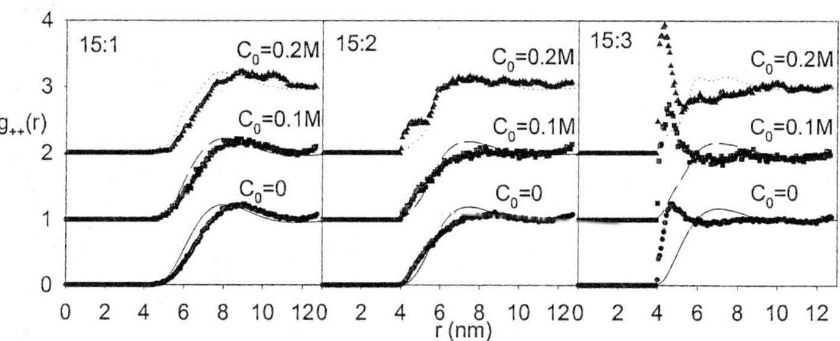

Figure 1. Macroion-macroion PCFs. Symbols and lines represent MC data and DFT results respectively. The curves for C_0=0.1M and 0.2M have been consecutively shifted upward by one unit.

Figures 1 to 6 display the calculated results from DFT and Monte Carlo simulations for the PCFs under a number of solution conditions. In these calculations, the concentration of neutral component C_0 varies from 0 to 0.2M, the valence of counterions Z_- varies from -1 to -3, and the total volume occupation of neutral particles varies from 4.03% to 29.86%. In Figures 1 and 2, one may observe that the repulsion between macroions is reduced while that between counterions is increased with the increase of the counterion valence. The potential of mean force between similarly charged macroions becomes strongly attractive in trivalent solutions, in particular at high concentration of neutral species, while they are purely repulsive in the corresponding monovalent solution. In general, the correlation between like-charged ions decreases when more neutral particles are added. Figure 3 shows that the accumulation between unlike-charged ions is more evident with the increase of counterion valence or the content of neutral component. Figures 4 to 6 indicate that the change of counterion valence has only minor influence on the distributions of macroions, counterions and neutral particles near the surface of a fixed neutral particle but their accumulation in the vicinity of the fixed neutral particle will be enhanced when the concentration of neutral species rises.

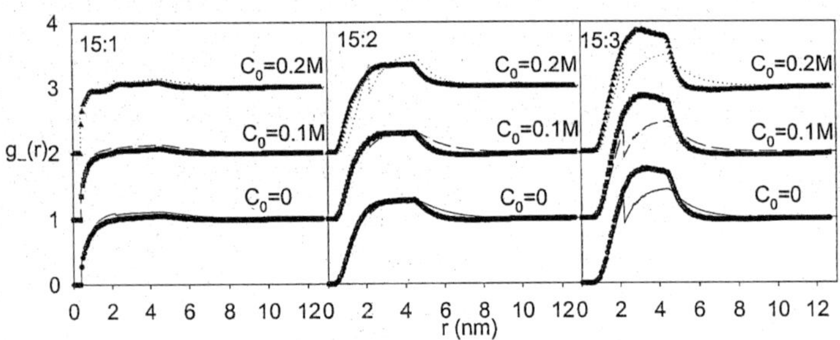

Figure 2. Counterion-counterion PCFs. Notation is the same as that in Figure 1.

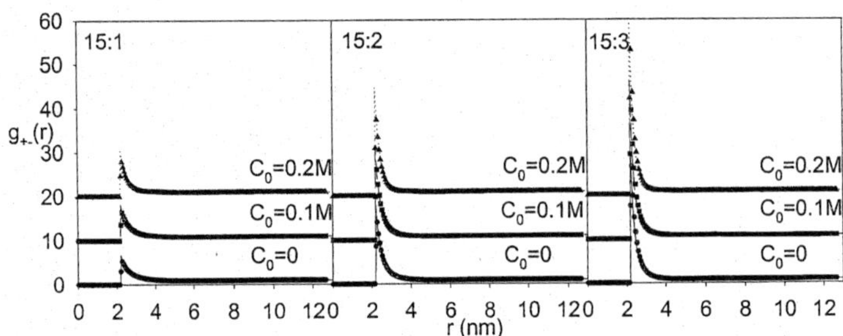

Figure 3. Macroion-counterion PCFs. The curves for C_0=0.1M and 0.2M have been consecutively shifted upward by ten units.

Under most circumstances, the DFT agrees fairly well with MC for predicting the macroion-counterion, neutral-macroion, neutral-counterion and neutral-neutral distributions. However, for systems with strong electrostatic interactions, i.e., in 15:2 and 15:3 solutions, the DFT fails to faithfully reproduce the macroion-macroion and counterion-counterion correlation functions. Besides a serious deviation on the macroion-macroion distributions, the DFT mistakenly predicts a discontinuity of the counterion-counterion pair correlation function. This deficiency is probably caused by the inaccuracy of the DCFs from MSA and the quadratic Talyor expansion for the intermolecular correlations.

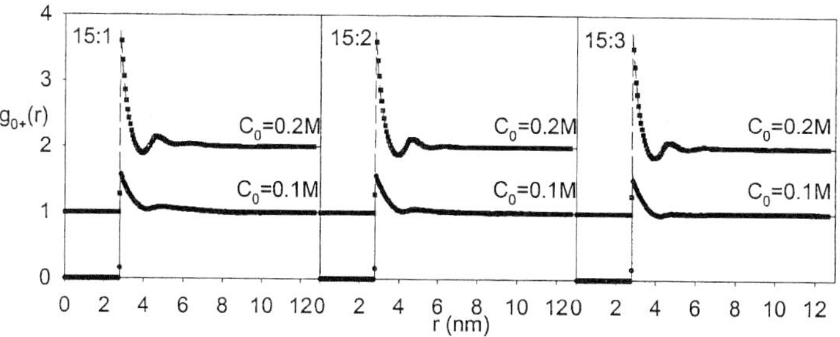

Figure 4. Neutral-macroion PCFs. The curves for C_0=0.2M have been consecutively shifted upward by one unit.

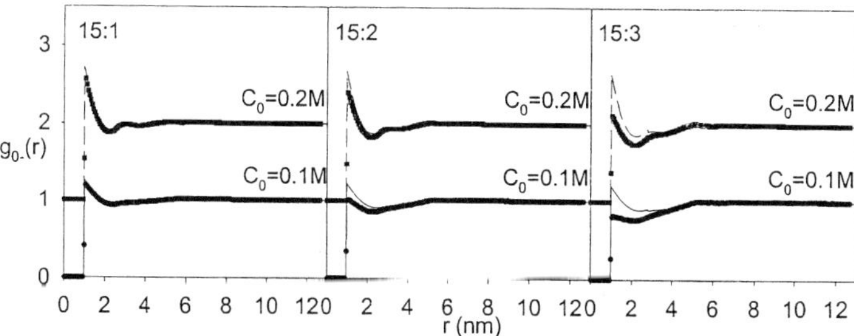

Figure 5. Neutral-counterion PCFs. Notation is the same as that in Figure 4.

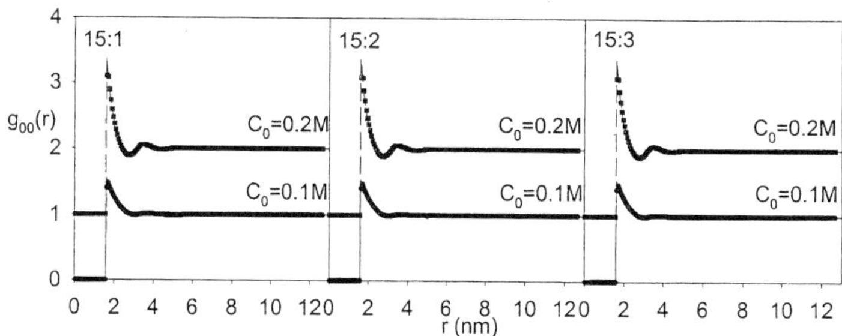

Figure 6. Neutral-neutral PCFs. Notation is the same as that in Figure 4.

Next we compare the thermodynamic properties of above systems calculated from DFT and MC. Table 1 gives the reduced excess internal energy per particle and the osmotic coefficients from MC and DFT. The agreement between theory and simulation is excellent for all cases studied in this work, even for systems in which DFT predicts inaccurate microscopic structures. This is because the thermodynamic properties are mainly determined by the macroion-counterion PCFs, which the DCF provides an accurate representation.

Table 1. The excess internal energies and osmotic coefficients obtained by MC and DFT.

		-Excess Internal Energy		**Osmotic Coefficient**	
C_0 (M)	Z_-	MC	DFT	MC	DFT
0	-1	1.70	1.78	0.741	0.747
	-2	4.13	4.31	0.429	0.421
	-3	7.12	7.11	0.0860	0.0857
0.1	-1	0.422	0.444	1.77	1.77
	-2	0.611	0.642	1.81	1.80
	-3	0.776	0.777	1.81	1.81
0.2	-1	0.245	0.261	3.50	3.49
	-2	0.497	0.355	3.53	3.59
	-3	0.416	0.420	3.61	3.62

Conclusion

We have shown that the DFT is able to describe the nonideality of highly asymmetric electrolyte and neutral component mixtures that simulate 'macromolecular crowding'. The PCFs, excess internal energies and osmotic coefficients predicted by DFT are in good agreement with results from MC simulations except when there is a very strong electrostatic interaction. Although the DFT fails to reproduce the macroion-macroion and counterion-counterion PCFs for systems with strong electrostatic interactions owing to the limitations of MSA and quadratic expansion, it successfully reproduces other PCFs and the thermodynamic properties.

Certainly the primitive model is oversimplified for representing 'macromolecular crowding', but it provides at least a step forward beyond merely excluded-volume considerations. Other interactions such as chain connectivity and van der Waals attractions also affect the behaviors of biological crowded environments. In our previous work, we have demonstrated the versatility of DFT to account for the influences of other non-bonded interactions such as chain connectivity, van der Waals attraction and association. [5-10] Therefore, the extention of the current DFT for crowding under more realistic situations will be a natural direction for our future work.

[1] R. J. Ellis and A. P. Minton, *Nature* **2003**, *425*, 27.
[2] A. P. Minton, *Biophys. Chem.* **1995**, *57*, 1.
[3] J. Rescic, V. Vlachy, L. B. Bhuiyan, et al., *Mol. Phys.* **1998**, *95*, 233.
[4] Z. D. Li and J. Z. Wu, *Phys. Rev. E* **2004**, in press.
[5] Y. X. Yu and J. Z. Wu, *J. Chem. Phys.* **2002**, *117*, 2368.
[6] Y. X. Yu and J. Z. Wu, *J. Chem. Phy.s* **2003**, *118*, 3835.
[7] C. Gu, G. H. Gao, and Y. X. Yu, *J. Chem. Phys.* **2003**, *119*, 488.
[8] Y. X. Yu and J. Z. Wu, *J. Chem. Phys.* **2002**, *116*, 7094.
[9] Y. P. Tang and J. Z. Wu, *J. Chem. Phys.* **2003**, *119*, 7388.
[10] Y. P. Tang and J. Z. Wu, *Phys. Rev. E* **2004**, *70*, 011201.

Stimulus Dependent Redistribution of Membrane Raft Cholesterol in Human Platelets

Kathleen Boesze-Battaglia, Richard J. Schimmel, Cheryl Gretzula*

Dept. Biochemistry, School of Dental Medicine, University of Pennsylvania, Philadelphia, PA 19104, USA

Summary: Cell membranes provide a requisite dynamic interface to facilitate communication between the extracellular environment and the intracellular milieu. These membranes contain proteins that span and/or are loosely associated with the lipid bilayer. The organization of lipids and proteins components into membrane micro-domains provides a temporal and spatial signaling platform for communication. Recently, cholesterol and sphingomyelin enriched membrane micro-domains known as lipid rafts have been implicated in cell signaling events. In these studies we have advanced our hypothesis that stimulus dependent rearrangement of cholesterol into and out of membrane rafts provides a unique lipid–mediated regulatory mechanism. Using fluorescent derivatives of cholesterol, we have shown that membrane raft associated cholesterol was altered in response to collagen-induced platelet aggregatory stimulation. Collagen stimulation resulted in a rapid redistribution of cholesterol from the outer to the inner membrane monolayer. The reorganization of the outer membrane monolayer resulted in a concomitant increase in outer monolayer fluidity. These studies are the first to show that membrane cholesterol was released from the exchangeable membrane raft pool in response to physiological stimuli.

Keywords: cholesterol; fluorescent probes; membrane; membrane raft; phospholipids

Introduction

Platelets are hematopoeitic cells essential in the maintenance of the vascular system. Through a series of sequential and coordinated mechanisms they are activated in response to vascular injury. In hypercholesterolemic individuals this activation process is attenuated due in part to the presence of high levels of serum cholesterol. The platelet plasma membrane is the site of interactions between extra cellular stimuli and the organized biochemical events necessary to support aggregation. It provides the charged phosphatidylserine rich surface for the activation of the prothrombinase complex as well as membrane raft signaling platforms. These cholesterol enriched platelet rafts were specifically enriched in the membrane glycoprotein CD36 (a

 DOI: 10.1002/masy.200550107

member of the tetraspanin protein family) and the surface protein GPIb, as well as Src p60 $^{c\text{-}src}$ and the Src-related kinases lyn; p53/56 ^{lyn} (1). This initial work suggested that platelet rafts were involved in platelet signal transduction. Recent work has provided more compelling data to support this hypothesis and moreover has met essential criteria in the characterization of detergent resistant membranes (DRMs) as membrane rafts which includes a detailed description of the lipid composition of the platelet DRMs (2). The phosphatidylinositol (PI) pool of phospholipids was the most significantly affected in DRMs isolated from thrombin-stimulated platelets. The change in DRM morphology in the thrombin stimulated platelets correlated with an increase in phosphatidic acid and products of the phosphoinositide-3 kinase reaction.

Although membrane rafts have been well characterized, a considerable amount of debate has focused on their relative distribution between the inner and outer membrane monolayer. We have previously documented a stimulus dependent redistribution of platelet cholesterol in response to the strong aggregatory stimuli, collagen. In these studies we provide evidence suggesting that this stimulus dependent reorganization is a translocation of cholesterol from cholesterol enriched outer monolayer raft to an inaccessible inner monolayer cholesterol pool. We hypothesize that this intracellular pool forms inner monolayer rafts and is required for platelet aggregation.

Materials and Methods

Materials. Triton X-100 was purchased from Sigma (St. Louis, MO). Octylglucopyranoside (OG) was purchased from Calbiochem. Cholesterol and phosphatidylcholine (PC) were purchased from Avanti Polar Lipids (Birmingham, Alabama).

Platelet Isolation and Additional Assays

Platelets were isolated from platelet rich plasma obtained from the American Red Cross. Erythrocytes were removed from the platelet-rich plasma by centrifugation at 750 rpm, the platelets were collected by centrifugation at 2500 rpms for 20 min. and the platelets resuspended in platelet buffer; 145mM NaCl, 5 mM KCl, 1mM MgSO4, and 10 mM Hepes, pH 7.4. This platelet suspension was filtered on a column of 40 ml Sepharose 4B (Pharmacia) to remove remaining plasma (3). Eluted platelets were centrifuged at 750 rpm to remove residual erythrocytes and suspended in platelet buffer to a final volume of 1-3 mls. Aliquots of the platelet suspension were removed for determination of cholesterol (4) and phospholipid-phosphate (5).

Incorporation of Fluorescent Probes

22-(N-(7-nitrobenz-2-oxa-1,3- diazol-4-yl)amino)-23,24-bisnor-5- cholen-3-ol (NBD-CHOL) and cholestatrienol (C-3) were incorporated into platelets using small unilamellar vesicles (SUVs) as a donor (6). Briefly, phosphatidycholine, cholesterol and NBD-cholesterol or cholestatrienol (probes added at 2 mole% relative to total lipid) were dissolved in 2:1 chloroform: methanol, dried under nitrogen, lyophilized to remove trace amounts of organic solvents, resuspended in platelet buffer and sonicated to clarity (Branson 500 Sonicator). Large unilamellar species were removed by centrifugation at 45,000 rev min-1 for 30 min (7). The SUV exchange vesicles contained 40 mole % cholesterol so as not to alter the endogenous cholesterol content of the platelets. Platelets were incubated with probe containing exchange vesicles for 3 h at 37°C (8) and recovered by centrifugation at 2500 rpms (15min). The fluorescently labeled platelet pellet was resuspended in 2-5mls platelet buffer and used immediately for membrane raft isolation as described below. For membrane fluidity studies platelets were labeled with cis-parinaric acid as described (Sklar, et al., 1970).

Fluorescence Measurements

All fluorescence measurements were performed on a Perkin-Elmer LS50B spectrofluorimeter as described (6). Prior to the start of each experiment $CaCl_2$ was added to the platelet suspension to a final concentration of 0.10 mM, platelets were transferred to cuvettes placed into a water-jacketed turret at 37°C. The intra-membrane distribution of NBD-cholesterol was determined using dithionate dependent fluorescence quenching of this probe as described (9) and of C-3 using fluorescence resonance energy transfer upon the addition of 2,4,6-trinitrobenznesulfonic acid as described (6).

Isolation of Triton X-100 Resistant Membrane Rafts

Triton X-100 resistant membrane fractions were prepared from platelet membranes essentially as described (10). Freshly isolated platelet membranes were suspended in MOPS buffer to a final protein concentration of 8-10 mg/ml. Triton X-100 (2% w/v) was added to this suspension to a final concentration of 1% w/v Triton X-100 and the suspensions were homogenized with three passes of a glass pestle through glass Tenboeck Tissue Grinder. Homogenates were prepared in the light or in the dark. The homogenate was mixed with 1.23 ml of 2.4 M sucrose to yield a final sucrose concentration of 0.9 M and transferred to a SW-41 centrifuge tube. The sample was overlaid with sucrose solutions in MOPS buffer of decreasing concentrations of 0.8M, 0.7M, 0.6M and 0.5M sucrose and centrifuged at 46,000 rpm for 20 hours at 4°C. The fractions were collected and analyzed for cholesterol, phosphate and protein. In control experiments equal volumes of platelets were treated with 2% octylglucopyranoside (OG) as described (11). The isolated rafts were analyzed for total ganglioside levels by quantitative immunoblotting using anti-cholera toxin antibody as described (12).

Cholesterol Oxidase Treatment

Platelets were incubated with cholesterol oxidase (*Nocardia.sp*, Calbiochem, La Jolla, CA) at a concentration of 5 units/ml platelets. The platelets were treated with collagen and cholesterol oxidase simultaneously, and aliquots removed at the indicated times. The reaction was terminated with the addition of ice-cold methanol, and lipids immediately extracted into chloroform (13). The extracted lipids were analyzed for cholesterol and cholestanone content by HPLC as described (6).

Results

Cholestatrienol Associates with Platelet Membrane Rafts

Cholesterol enriched platelet rafts provide a signaling matrix necessary to promote aggregatory responses. Although rafts have been rigidly defined based on phase behavior as lipids in the liquid ordered phase, in our studies, rafts are more loosely defined as low buoyant density detergent resistant membrane fractions. In these studies we sought to identify a fluorescent cholesterol marker that would reflect membrane raft cholesterol behavior. We focused on two well-characterized cholesterol derivatives, NBD-cholesterol and cholestatrienol. Cholestatrienol is structurally most similar to cholesterol and is efficiently incorporated into both cell membranes and unilamellar vesicles. It distributes into the bilayer and has exchange rates similar to those obtained using [^{3}H] cholesterol (14). In contrast NBD-cholesterol has a larger bulky NBD moiety, which led us to hypothesize that this probe most likely reflects the behavior of cholesterol-poor non-raft domains. Platelets were isolated and labeled with one of these two probes as described above. Labeled platelets were treated with ice-cold Triton X-100 and membrane rafts isolated. As shown in Figure 1, fractions 3-5 contained a low buoyant density membrane species enriched in cholesterol and ganglioside 1 (GM1), indicative of a membrane raft. In contrast, the Triton X-100 soluble species were isolated in fractions 12-14 as expected based on previous studies (15); (2). In order to determine if the cholesterol probes were reversibly associated with the raft fraction all fractions were analyzed for NBD-cholesterol and cholestratrienol. Cholestatrienol was preferentially associated with the low buoyant density membrane raft fraction (fractions 3-5). In contrast NBD-cholesterol was found in the non-raft Triton X-100 soluble fractions (fractions 12-14). These studies suggest that cholestatrienol can be used as a marker of membrane raft cholesterol behavior and NBD-cholesterol as a marker for non-raft cholesterol. Moreover, cholestatrienol can be used to probe the dynamic properties of platelet membrane raft cholesterol pools.

Cholestatrienol Accessibility to Quenching Decreases upon Collagen Stimulation of Platelets

We have previously shown that platelet membrane cholesterol becomes inaccessible to outer monolayer quenching reagents upon stimulation with collagen (6). Those early studies suggested that cholesterol translocates from the outer to the inner membrane monolayer. In the next part of this study we wanted to determine if cholesterol translocates into or out of an outer monolayer membrane raft. Using C-3 and NBD-cholesterol to follow movement of membrane cholesterol in raft and non-raft micro-domains we asked which of these two probes becomes less accessible to outer monolayer quenching upon stimulation with collagen. Platelets were labeled with either C-3 or NBD-cholesterol and outer monolayer associated probe quantitated using 2,4,6-trinitrobenezenesulfonic acid (TNBS) or dithionate, respectively (6). We have previously shown that these probes incorporate into the platelet plasma membrane with a $t_{1/2} = 39$ min and -k= -0.0175 fluorescence units/ min. As seen in Figure 2, when C-3 labeled platelets were stimulated with collagen (10ug/ml), a decrease in C-3 accessibility was observed

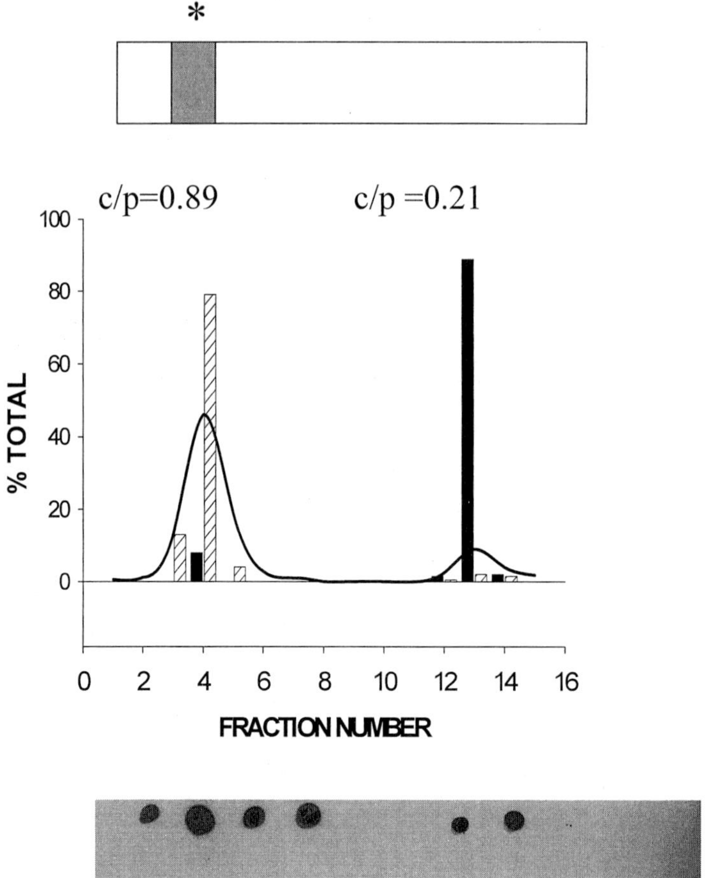

Fig. 1. Cholestatrienol (C-3) preferentially partitions into platelet membrane rafts. Freshly isolated human platelets labeled with either C-3 or NBD-cholesterol were treated with ice-cold Triton X-100 and the detergent resistant membrane fractions isolated (10). The low buoyant density fraction is indicted in gray and the % of the total membrane cholesterol is indicated on the ordinate and the GM1 levels shown on the dot blot. The cholesterol to phospholipid mole ratio (c/p) of the two major fractions; the membrane raft, fractions 3-5, and the Triton soluble fractions, 12-14 are indicated in the upper panel. The amount of NDB-cholesterol and cholestatrienol in each fraction is indicated.

within the first 5 minutes; no detectable change in NBD-cholesterol accessibility to probe was observed over this same time frame. Over the course of 30 minutes, the amount of C-3 accessible to quenching decreased from 60% to approximately 40%, with the most rapid translocation occurring within the first 5 minutes (Figure 3). In addition, there is no apparent redistribution of C-3 (from inner to outer monolayer) once it becomes inaccessible, as indicated by no change in accessibility for up to 30 minutes. Since C-3 preferentially labeled the membrane raft associated pool these results suggest that C-3 translocates out of a membrane raft upon collagen stimulation. We correlated these changes in C-3 accessibility with the translocation of phosphatidylserine from the inner to the outer monolayer. When NBD-PS labeled platelets were treated with collagen, there was a redistribution of PS, such that within the first five minutes there was a 50% increase in outer monolayer accessible PS (data not shown). Thus suggesting that cholesterol redistribution occurs on the same time scale as PS translocation; a necessary step in the platelet aggregator response.

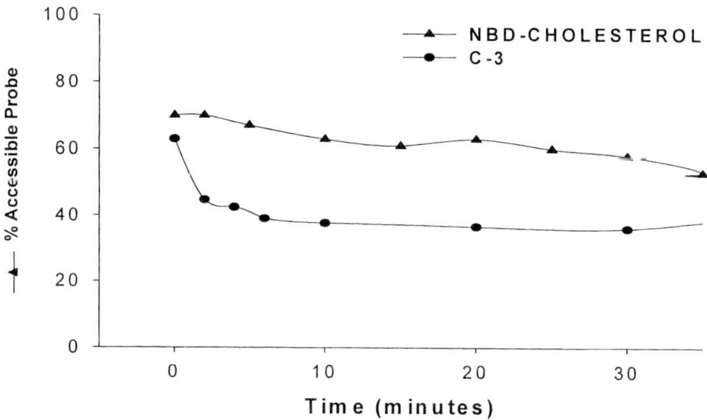

Fig. 2. C-3 Redistributes within the platelet plasma membrane upon collagen stimulation. Platelets labeled with either C-3 or NBD –cholesterol were stimulated with collagen (10 ug/ml) and aliquots removed at the times indicted. The amount of C-3 or NBD-cholesterol accessible to quenching by TNBS or dithionite respectively is indicated on the ordinate.

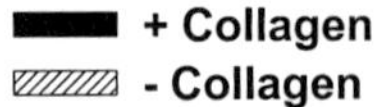

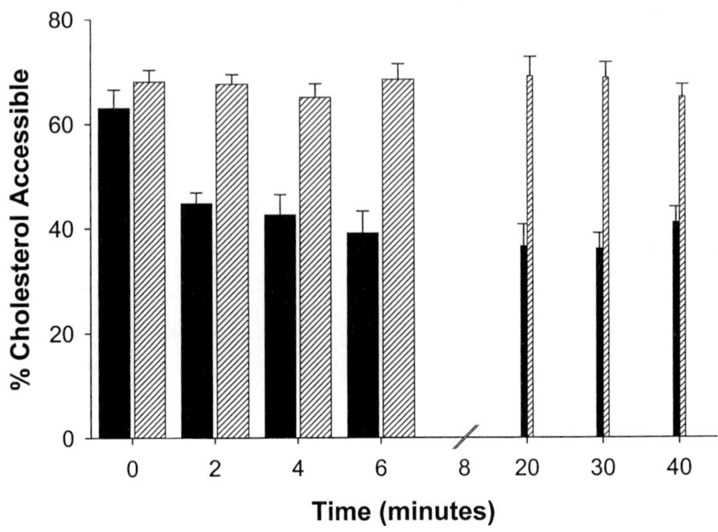

Fig. 3. Collagen decreases the accessibility of cholestatrieonol to TNBS quenching within the first five minutes. Platelets were labeled with C-3 and incubated in the presence [filled column] or absence [striped column] of collagen 10ug/ml for the times indicated on the abscissa. The amount of C-3 in the outer leaflet is determined as the decrease in cell-associated fluorescence after quenching with TNBS and calculated as described (Boesze-Battaglia, et al. 1996). Each value presented is the mean and standard error of four replicate experiments for platelets incubated in the absence of collagen and six experiments for platelets incubated with collagen. Each experiment was performed on an individual platelet preparation.

Stimulation With Collagen Alters The Distribution of Endogenous Platelet Membrane Cholesterol and Outer Monolayer Fluidity

The decrease in accessibility of C-3 to quenching upon collagen stimulation suggests that this probe translocates from the outer to the inner membrane monolayer. While the use of fluorescent sterol probes to study membrane cholesterol distribution has been well documented (16); (17), a caveat in the present studies is that the C-3 experiments do not directly study the distribution of endogenous cholesterol. Outer monolayer cholesterol is susceptible to oxidation by cholesterol oxidase (18);(19);(20). A product of this oxidation reaction, cholest-4-en-3-one (cholestanone), is readily quantitated by HPLC (6). In this study we asked if endogenous cholesterol was more or less susceptible to oxidation by cholesterol oxidase upon stimulation with collagen. Platelets were treated with collagen and subsequently with cholesterol oxidase as described in methods. If endogenous cholesterol redistributes from the outer to the inner membrane monolayer we anticipate a decrease in oxdizable cholesterol reflected as a decrease in cholestanone. As seen in Figure 4, the amount of oxidizable cholesterol decreased by almost half upon stimulation of platelets with collagen.

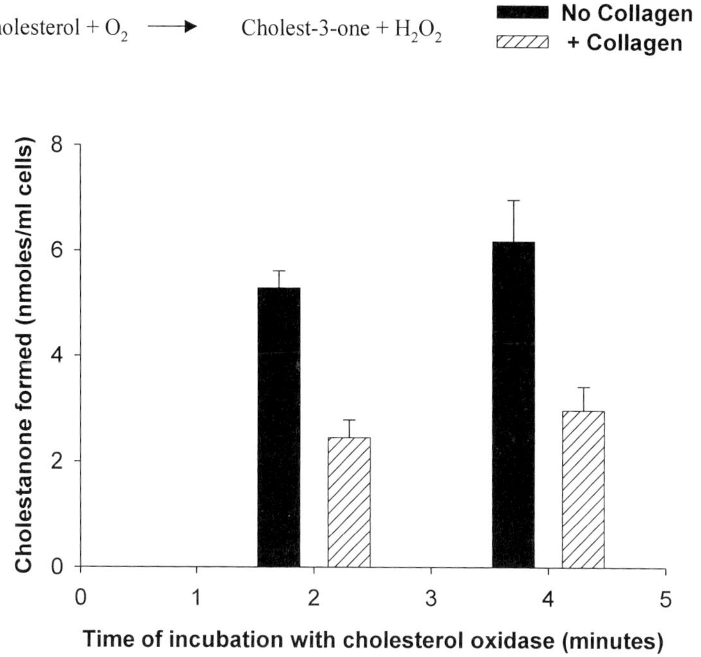

Fig. 4. Collagen treatment decreases the enzymatic conversion of endogenous cholesterol to cholestenone. Platelets were incubated in the presence or absence of collagen (10ug/ml) and cholesterol oxidase (5U/ml platelets) for the times indicated. The reaction was terminated and the cholestenone content determined by HPLC. Each value presented is the mean and standard error of eight observations.

The time course of this decrease; changes occurred within the first 5 minutes was identical to that observed with C-3 (Figure 3). We have previously shown that the redistribution of endogenous cholesterol is not directly dependent on ATP since pre-incubation of platelets with iodoacetamide, resulting in a decrease in ATP by 50 %, had no effect on cholesterol oxidase in stimulated platelets as described in Methods (6).

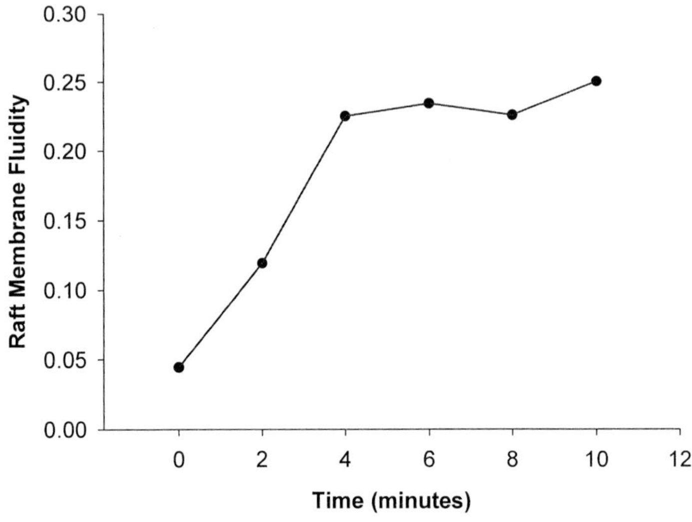

Fig. 5. Collagen treatment increases the fluidity of the outer membrane monolayer. Platelets were labeled with cis-parinaric acid treated with collagen (10ug/ml) for the times indicated and the fluorescence polarization of the cis-parinaric acid determined at 37°C. The results are representative of a series of five individual platelet preparations.

Lastly, to test the hypothesis that the movement of C-3 and endogenous cholesterol was from the outer to the inner membrane monolayer as opposed to a lateral diffusion within the plane of the outer monolayer, we measured the dynamic properties of the outer membrane monolayer. The fluorescent probe cis-parinaric acid has been used to measure outer membrane monolayer fluidity. Platelet membranes were labeled with cis-parinaric acid, and the change in outer monolayer fluidity measured as fluorescence polarization was monitored over time after collagen stimulation. As seen in figure 5, there was a rapid increase from 0.05 to 0.25 polarization units over the first 5 minute. This is the same time frame in which a redistribution of both C-3 and endogenous cholesterol was observed. Collectively, these results suggest that collagen stimulates a translocation of cholesterol from an accessible (outer monolayer) to an inaccessible (inner monolayer) pool. Moreover, movement is most likely out of a cholesterol enriched outer monolayer raft and directly alters the fluidity of the outer leaflet.

Discussion

We have previously documented an intramembrane redistribution of cholesterol upon stimulation of human platelets with collagen. In this study we employed fluorescent sterol analogs to probe this redistribution of cholesterol as raft or non-raft associated in response to platelet stimulation by collagen. We have shown that C-3, a membrane raft specific cholesterol probe, as well as endogenous cholesterol, redistributes in platelet membranes upon collagen stimulation. Although the present studies do not directly distinguish between a lateral redistribution or the movement of cholesterol from the outer to the inner membrane monolayer, indirect evidence for the later comes from a change in membrane fluidity as the cholesterol translocates. Using cis-parinaric acid to probe the change in outer membrane monolayer dynamics (Figure 5), we showed that upon collagen stimulation and cholesterol movement the fluidity of the outer membrane monolayer increased. It is well documented that such increases in fluidity are often due to a decrease in localized cholesterol levels. Thus, the fluidity results are consistent with redistribution of cholesterol to the inner monolayer.

Although we favor the view that cholesterol translocates out of the membrane raft pool, it is not inconceivable to hypothesize that cholesterol redistributes to a collagen-membrane interaction zone that is inaccessible to quenching and cholesterol oxidase treatment. Although a viable hypothesis, one consistent with a lateral reorganization of cholesterol in rafts as proposed by Brodin, et al., (23), a number of experimental observations argue against this interpretation. The redistribution of cholesterol in response to stimulation is not unique to collagen as described here but is also seen in response to stimulation by ADP (21). Since collagen and ADP act through two different receptor classes; G-protein coupled receptors and the GpIIb/IIIa family of integrin like receptors we propose that the translocation of cholesterol is not due to the formation of a collagen-membrane reaction zone which is inaccessible to quenching reagents. In addition, TNBS is a small water soluble quenching agent and is not predicted to have limited access to C-3. The most compelling support for a translocation of cholesterol out of the membrane raft comes from cholesterol enrichment studies. In those studies when platelet membrane cholesterol is enhanced the C-3 accessibility profile is virtually identical to that observed upon collagen treatment, suggesting that the added cholesterol is incorporated into an inaccessible possibly inner membrane raft pool (unpublished observations, Boesze-Battaglia and Schimmel).

Cholesterol alters platelet response to aggregatory stimuli in vitro (22); (23) and *in vivo* hypercholesterolemia results in abnormal clotting and the formation of atherosclerotic plaque. The precise role of cholesterol in the aggregatory response is under intense investigation. A number of signaling proteins and lipid second messengers have been found in membrane rafts of both resting and stimulated platelets (23). In recent studies, Gousset, et al., (15), have suggested that raft formation is a dynamic reversible event triggered by activation, they however provide little evidence to confirm the reversible nature of this domain formation. We propose a variation on this theme; that upon stimulation, cholesterol redistributes out of an outer monolayer membrane raft into an inaccessible- inner monolayer pool of cholesterol. We propose that this monolayer pool is a transient raft complex involved in organizing lipid mediators to further platelet aggregation and commit the platelet to an irreversible aggergatory event. Because it has been well documented that some platelet agonists mediate reversible platelet aggregation while stronger agonist mediate irreversible aggregation and rapid phospholipid translocation, it will be intriguing to determine how membrane raft formation is correlated with platelet phospholipid translocation and the reversible versus irreversible processes. Collectively this very new and growing body of work coupled with the stimulus dependent redistribution of PS, PE and cholesterol provides an ideal model system in which to study membrane raft formation and dissolution in response to stimuli. The dynamic redistribution of cholesterol might represent a critical mechanism for the early steps of platelet activation under physiological conditions (2); (24). In a manner analogous to hematopoietic cells, small lipid rafts may cluster into larger raft signaling platforms- traversing both monolayers of the plasma membrane bilayer. Moreover, the stimulus dependent reorganization of cholesterol provides an attractive mechanism to explain the membrane structural aspects of shape change. Shape change and the formation of filipodia (in platelets) and matrix vesicles (in chondrocytes and osteoblasts) requires areas with small radii of curvature, a structure more readily achieved in a cholesterol-poor region of the membrane.

[1] Dorahy, D. J., L. F. Lincz, C. J. Meldrum, and G. F. Burns. 1996. Biochemical isolation of a membrane microdomain from resting platelets highly enriched in the plasma membrane glycoprotein CD36. *Biochem. J. 319:67.*

[2] Bodin, S., S. Giuriato, J. Ragab, B. M. Humbel, C. Viala, C. Vieu, H. Chap, and B. Payrastre. 2001. Production of phosphatidylinositol 3,4,5-triphosphate and phosphatidic acid in platelet rafts: Evidence for a critical role of cholesterol-enriched domains in human platelet activation. *Biochemistry 40(50):15290.*

[3] Carvalho, A. C. A., R. W. Colman, and M. Lees. 1974. *N. Engl. J. Med. 290:434.*

[4] Allain, C. C., L. S. Poon, C. S. Chan, W. Richmond, and P. C. Fu. 1974. Enzymatic determination of total serum cholesterol. *Clinical Chemistry 20:470.*

[5] Bartlett, G. R. 1959. Phosphorous assay in column chromatography. *J. Biol. Chem. 234:466.*

[6] Boesze-Battaglia, K., S. T. Clayton, and R. J. Schimmel. 1996. Cholesterol redistribution within human platelet plasma membrane: Evidence for a stimulus-dependant event. *Biochemistry 35:6664.*

[7] Barenholtz, Y., D. Gibbs, B. J. Litman, T. Thompson, and F. D. Carlson. 1977. *Biochemistry 16:2806.*

[8] House, K., D. Badgett, and A. Albert. 1989. *Exp. Eye Res. 49:561.*

[9] Mc Intyre, J. C., and R. G. Sleight. 1991. Fluorescence assay for phospholipid membrane asymmetry. *Biochemistry 30:11819.*

[10] Boesze-Battaglia, K., J. Dispoto, and M. A. Kahoe. 2002. Association of a photoreceptor-specific tetraspanin protein, ROM-1, with triton X-100-resistant membrane rafts from rod outer segment disk membranes. *J. of Biological Chemistry 277:41843.*

[11] Nusrat, A., C. A. Parkos, P. Verkade, C. S. Foley, T. W. Liang, W. Innis-Whitehouse, K. K. Eastburn, and J. L. Madara. 2000. Tight junctions are membrane microdomains. *J. Cell Sci. 113:1771.*

[12] Millan, J., M. Qaidi, and M. A. Alonso. 2001. Segregation of co-stimulatory components into specific T cell surface lipid rafts. *Eur. J. Immunol. 31:467.*

[13] Folch, J., M. Lees, and M. G. A. Sloane-Stanley. 1957. *J. Biol. Chem. 226:497.*

[14] Yeagle, P. L., A. Albert, K. Boesze-Battaglia, J. E. Young, and J. Frye. 1989. Cholesterol Dynamics in Membranes. *Biophysical J. 57:413.*

[15] Gousett, K., W. Wolkers, N. M. Tsvetkova, A. E. Oliver, C. Field, N. J. Walker, J. H. Crowe, and F. Tablin. 2002. Evidence for a Physiological Role for Membrane Rafts in Human Platelets. *J. Cellular Physiology 190: 117.*

[16] Schroeder, F., J. R. Jefferson, A. B. Kier, J. Knittel, T. Scallen, Wood, W.G., and I. Hapala. 1991. *Proc. Soc. Exp. Biol. Med. 196:235.*

[17] Schroeder, F., A. Gallegos, M., B. P. Atshaves, S. M. Storey, A. L. McIntosh, A. D. Petrescu, H. Huang, O. Starodub, H. Chao, H. Yang, A. Frolov, and A. B. Kier. 2001. Recent Advances in Membrane Microdomains: Rafts, Caveolae, and Intracellular Cholesterol Trafficking. *Exp. Biol. Med. 226:873.*

[18] Lange, Y. 1992. *J. Lipid Res. 33:315.*

[19] Dawidowitz, E. A. 1987. *Current Top. Membrane Transp. 29:175.*

[20] Brasaemle, D. L., and A. D. Attie. 1990. *J. Lipid Res. 31:103.*

[21] Boesze-Battaglia, K., and R. J. Schimmel. 1999. Collagen-stimulated unidirectional translocation of cholesterol in human platelet membranes. *J. Exp. Biol. 202:453.*

[22] Shattil, S. J., R. Anaya-Galindo, J. Bennett, R. W. Colman, and R. A. Cooper. 1975. Platelet hypersensitivity induced by cholesterol incorporation. *J. Clin. Invest. 55:636.*

[23] Schimmel, R. J., P. Shah, M. Fenner, and K. Boesze-Battaglia. 1997. Cholesterol enhances the adhesion of human platelets to fibrinogen: studies using a novel fluoresence based assay. *Platelets 8:216.*

[24] Bodin, S., H. Tronchere, and B. Payrastre. 2003. Lipid rafts are critical membrane domains in blood platelet activation processes. *Biochim. Biophys. Acta 1610:247.*

Interaction of Polyunsaturated Fatty Acids with Cholesterol: A Role in Lipid Raft Phase Separation

Stephen R. Wassall,[1] Saame Raza Shaikh,[2] Michael R. Brzustowicz,[1] Vadim Cherezov,[3] Rafat A. Siddiqui,[4] Martin Caffrey,[3] William Stillwell[2]*

[1]Department of Physics, Indiana University Purdue University Indianapolis, 402 N. Blackford Street, Indianapolis, IN 46202-3273, USA
[2]Department of Biology, Indiana University Purdue University Indianapolis, 723 W. Michigan Street, Indianapolis, IN 46202-5132, USA
[3]Biochemistry, Biophysics, and Chemistry, The Ohio State University, Columbus, OH 43210-1173, USA
[4]Cellular Biochemistry Laboratory, Methodist Research Institute at Clarion Health, 1701 N. Senate Ave., Indianapolis, IN 46202, USA

Summary: Unequal affinity between lipids has been hypothesized to be a mechanism for the formation of microdomains/rafts in membranes. Our studies focus upon the interaction of cholesterol with polyunsaturated fatty acid (PUFA)-containing phospholipids. They support the proposal that steric incompatibility of the rigid steroid moiety for highly disordered PUFA chains, in particular docosahexaenoic acid (DHA), provides a sensitive trigger for lateral segregation of lipids into PUFA-rich/sterol-poor and PUFA-poor/sterol-rich regions. Solid state ^{2}H NMR and x-ray diffraction (XRD) demonstrate that the solubility of cholesterol is reduced in 1-palmitoyl-2-docosahexaenoyl-phosphatidylethanolamine (16-0:22:6PE) bilayers. In mixed membranes of phosphatidylethanolamine (PE) with the lipid raft forming molecules egg sphingomyelin (SM) and cholesterol, diminished affinity of the sterol for 16:0-22:6PE relative to 1-palmitoyl-2-oleoylphosphatidylethanolamine (16:0-18:1PE) is identified by ^{2}H NMR order parameters and detergent extraction. Phase separation of the PUFA-containing phospholipid from SM/cholesterol rafts is the implication, which may be associated with the myriad of health benefits of dietary DHA.

Keywords: cholesterol; docosahexaenoic acid (DHA); lipid domains; polyunsaturated fatty acid (PUFA)

Introduction

There is tremendous interest in understanding the properties of membranes containing polyunsaturated fatty acids (PUFA), especially docosahexaenoic acid (DHA).[1] This PUFA belongs to the ω3 series in which the ultimate unsaturation is 3 carbons from the terminal methyl or ω end and, with 22 carbons and 6 double bonds, constitutes the most unsaturated fatty acid commonly found in nature. A major issue that motivates the research is the enormous range of health benefits including the prevention of cancer and heart disease that accompany consumption of DHA found in fish oils. The diversity of the

 DOI: 10.1002/masy.200550108

physiological processes affected suggests a fundamental mode of action common to most cells. Our experiments target the plasma membrane as a likely site of action and focus upon the interaction of cholesterol with DHA-containing phospholipids. They support the idea that steric incompatibility of the rigid steroid moiety for highly disordered PUFA chains promotes lateral segregation of lipids into DHA-rich/sterol-poor and DHA-poor/sterol-rich regions, such as lipid rafts that are enriched in cholesterol and sphingolipids.[2,3] We speculate that concomitant changes in lipid raft-mediated cellular signaling events may be responsible in part for the dietary advantage conferred by DHA.

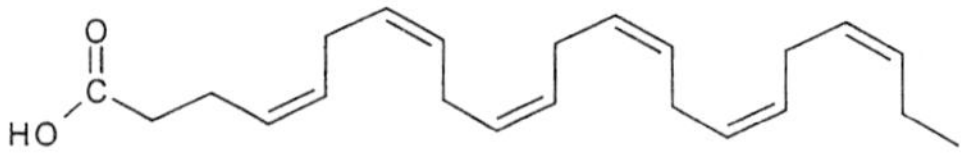

Structure of DHA

A flexible structure with rapid inter-conversion between torsional states is revealed for DHA chains esterified to phospholipids by molecular dynamics (MD) simulations.[4-5] The low energy barrier to rotational isomerization about the pairs of single C-C bonds that separate the unsaturated carbon atoms more than offsets the rigidity of the multiple double bonds. It is the high degree of conformational flexibility that distinguishes the molecular organization of PUFA from saturated and less unsaturated fatty acids. Besides producing a disordered membrane interior[6-8], the impact of DHA upon membranes includes increased permeability[9], phospholipid flip-flop[10] and in-plane elasticity.[11] The high conformational disorder of the DHA chain also prevents the close approach from the rigid steroid moiety of cholesterol that is allowed by the predominantly all-trans configuration of the upper portion of saturated chains. Our measurement of a factor of 3-4 reduction in the solubility of cholesterol in membranes composed of PC (phosphatidylcholine) with DHA or arachidonic acid at both sn-1 and-2 positions compared to more typical PC's with a saturated sn-1 chain exemplifies the poor affinity of the sterol for PUFA.[12-14] Partition coefficients (K_A^B) measured for cholesterol in unilamellar vesicles corroborate.[15]

Biological membranes are no longer considered to be a homogeneous mixture of lipid and protein but are envisaged as a complex arrangement of domains of differing composition.[16] Unequal affinity between different lipid species or between lipids and membrane proteins is responsible for the formation of these patches. Differential interaction with cholesterol provides an influential lipid-driven mechanism whereby sterol-rich/liquid ordered (l_o) and sterol-poor/liquid disordered (l_d) lipid domains segregate within lipid/cholesterol mixtures. Lipid rafts that are enriched in cholesterol and saturated sphingolipids and serve as platforms for signaling proteins, in particular, have recently attracted considerable interest. The formation of these domains reflects a preference of the sterol for sphingolipids. Their size, typically quoted ~ 50 nm, is the subject of debate.

The poor affinity of PUFA for cholesterol has been hypothesized to drive lateral phase separation into PUFA-poor/sterol-rich and PUFA-rich/sterol-poor microdomains.[12-14,17,18] We extend this hypothesis to a possible explanation of the health benefits of DHA. In most human cells DHA levels are low (typically < 5 % of the total acyl chains) and dietary supplementation primarily incorporates DHA into the sn-2 position of membrane phospholipids, often phosphatidylethanolamine (PE), while the sn-1 chain remains saturated.[19] We propose that DHA-containing phospholipids will be excluded from l_o sphingomyelin-rich/sterol-rich lipid rafts and will locate as DHA-rich microdomains within the l_d remainder of the membrane.[2,3] The partitioning into these regions of proteins that require a l_o vs. l_d environment to function will then be promoted and cellular signaling altered.

Here we present evidence in support of our proposal from predominantly solid state ^{2}H NMR spectroscopy supplemented by x-ray diffraction (XRD). The data amassed on 1-palmitoyl-2-docosahexaenoylphosphatidylethanolamine (16:0-22:6PE) demonstrate that this DHA-containing phospholpipid has low affinity for cholesterol and suggest that it separates from lipid raft molecules in mixed membranes.

Interaction of a DHA-Containing Phospholipid with Cholesterol

DHA Prohibits Close Contact with Cholesterol

Figure 1 shows solid state ^{2}H NMR spectra that are representative of the phases formed by [$^2H_{31}$]16:0-22:6PE in aqueous dispersion without and with equimolar cholesterol. Each spectrum is a superposition of powder patterns from all positions along the perdeuterated [$^2H_{31}$]16:0 sn-1 chain.[20] At -5 °C, the spectrum for [$^2H_{31}$]16:0-22:6PE is a broad featureless pattern with edges at ±63 kHz that reflects the slow reorientational motion undergone by acyl chains in the lamellar gel phase (Fig. 1a). The melting of the chains that accompanies the gel to liquid crystalline phase transition is apparent in the much narrower spectrum at 7.5 °C (Fig. 1 b). A plateau region of almost constant order in the upper part of the [$^2H_{31}$]16:0 sn-1 chain produces the well-defined edges at ±17 kHz. Increasingly more disordered methylenes in the lower portion of the chain give rise to the individual peaks, while the central pair of peaks is due to the highly mobile terminal methyl group. The spectrum for [$^2H_{31}$]16:0-22:6PE at 40 °C with edges at ±6 kHz is characteristic of the inverted hexagonal (H_{II}) phase (Fig. 1c).[21] Rapid diffusion of lipid molecules around the cylindrical structures that comprise this phase is responsible for an additional reduction by a factor of ½ in spectral width relative to the lamellar phase.

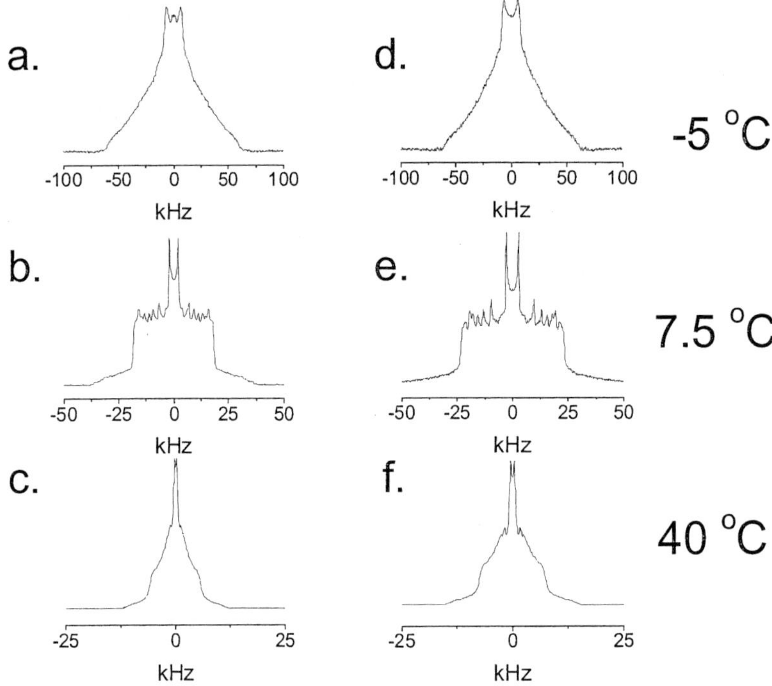

Figure 1. ^{2}H NMR spectra for 50 wt% aqueous dispersions in 50 mM Tris (pH 7.5) of a-c. [$^2H_{31}$]16:0-22:6PE and d-f. [$^2H_{31}$]16:0-22:6PE/cholesterol (1:1 mol).

The spectra for [$^2H_{31}$]16:0-22:6PE in the presence of equimolar cholesterol, remarkably, reveal the same phases. They are symptomatic of gel, liquid crystalline lamellar and H_{II} phases at –5 (Fig. 1d), 7.5 (Fig. 1e) and 40 °C (Fig. 1f), respectively. Apart from a slight narrowing at -5 °C and a broadening at 7.5 and 40 °C that are consistent with the sterol disrupting chain packing in the gel state and obstructing chain motion in the liquid crystalline phase, there is little perturbation to the spectra. Such a modest change is unusual. For most phospholipids the gel to liquid crystalline phase transition is broadened beyond detection with equimolar cholesterol.[22]

The variation with temperature of the first moments M_1 calculated from spectra that are plotted in Figure 2 elaborates upon the difference between the response to equimolar sterol of [$^2H_{31}$]16:0-22:6PE and [$^2H_{31}$]16:0-18:1PE. This quantity, defined by

$$M_1 = \frac{\int_{-\infty}^{\infty} |\omega| f(\omega) d\omega}{\int_{-\infty}^{\infty} f(\omega) d\omega} \qquad (1)$$

where $f(\omega)$ is the lineshape as a function of the frequency ω relative to the central Larmor frequency ω_0, is a sensitive indicator of membrane phase.[20] There are two discontinuities in the graph for [$^2H_{31}$]16:0-22:6PE (Fig. 2a). The transition from lamellar gel to liquid crystalline state gives rise to a sharp drop in the value of M_1 from $> 10 \times 10^4$ to $\sim 6.0 \times 10^4$ s^{-1} at 5 °C, and the transition to H_{II} phase results in a further drop to $< 3.0 \times 10^4$ s^{-1} centered at ~13 °C. Discontinuities associated with these phase changes, although somewhat modified, are still seen in the plot for [$^2H_{31}$]16:0-22:6PE/cholesterol (1:1 mol). In contrast, the M_1 values for our more typical [$^2H_{31}$]16:0-18:1PE control sample slowly decrease with temperature in the presence of cholesterol and do not exhibit the abrupt drop at 23 °C that in the absence of the sterol signifies the gel to liquid crystalline transition (Fig. 2b). We attribute the retention of a gel to liquid crystalline phase transition despite the addition of equimolar cholesterol to diminished solubility of the sterol in the DHA-containing phospholipid.

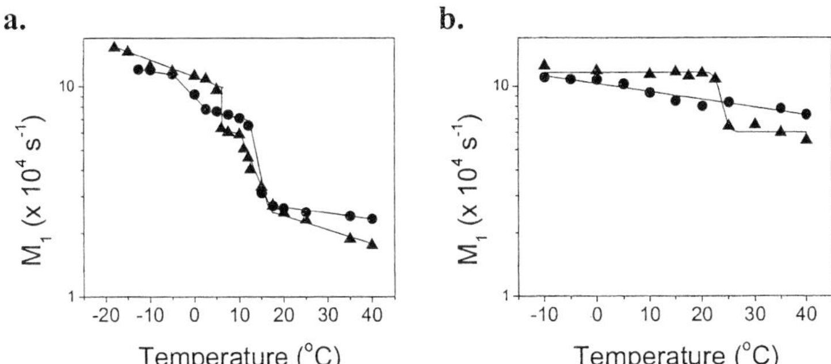

Figure 2. First moment M_1 vs. temperature for 50 wt% aqueous dispersions in 50 mM Tris (pH 7.5) of a. [$^2H_{31}$]16:0-22:6PE (▲) and [$^2H_{31}$]16:0-22:6PE/cholesterol (1:1 mol) (●), and b. [$^2H_{31}$]16:0-18:1PE (▲) and [$^2H_{31}$]16:0-18:1PE/cholesterol (1:1 mol) (●).

XRD confirms that the solubility of cholesterol is reduced in 16:0-22:6PE. Profiles of integrated radial intensity against reciprocal space ($q = 4\pi \sin\theta / \lambda$) for 16:0-

22:6PE/cholesterol and 16:0-18:1PE/cholesterol samples with a series of contents of added cholesterol χ_{chol} are presented in Figure 3. The method relies on detecting the second-order 002, 020 and 200 diffraction peaks from cholesterol monohydrate crystals that are formed outside the membrane when sterol in excess of the solubility limit is excluded.[13] Their respective reciprocal and real spacings are 0.3701 Å^{-1}, 17.0 Å (002); 1.033 Å^{-1}, 6.079 Å (020) and 1.044 Å^{-1}, 6.015 Å (200).[23] Inspection of Figure 3 reveals that sharp peaks (labeled ↓) from solid cholesterol are barely discernible at χ_{chol} = 32.5 mol% and subsequently grow in intensity as χ_{chol} increases for 16:0-22:6PE (left), whereas for 16:0-18:1PE they do not begin to emerge from the broad background due to membrane lipid until χ_{chol} exceeds >50 mol% (right).

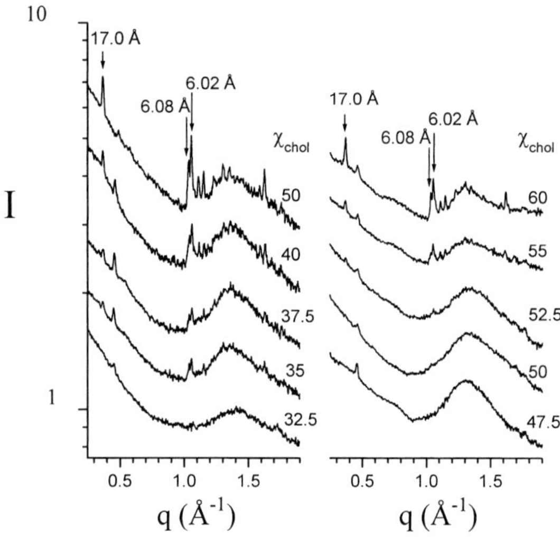

Figure 3. Integrated radial intensity profiles (I-q plots) for 50 wt% aqueous dispersions in 50 mM Tris (pH 7.5) of 16:0-22:6PE/cholesterol at 7.5 °C (left) and 16:0-18:1PE/cholesterol at 40 °C (right). The concentration of sterol χ_{chol} is listed in mol% to the right of each plot. At each temperature the systems are in the lamellar liquid crystalline state.

A precise measurement of the concentration at which cholesterol begins to precipitate from the membrane is obtained in Figure 4 by plotting the sum of the intensities of the three second-order reflections produced by solid sterol (normalized relative to the lipid wide-angle peak centered at $q \sim 1.4$ Å^{-1}) against χ_{chol}. Linear extrapolation to zero integrated intensity then gives a solubility of 32 ± 3 mol% for 16:0-22:6PE, substantially less than the value of 51 ± 2 mol% measured in 16:0-18:1PE that is comparable to previous estimates for many phospholipids.[24] Our explanation is that the highly disordered DHA chain is responsible for the smaller solubility of cholesterol in 16:0-22:6PE. This view may be reconciled with the "umbrella" model for cholesterol-phospholipid mixing proposed by Huang and Feigenson.[25] According to their model lipid head groups shield cholesterol from unfavorable contact with water and the solubility limit is reached when the head groups can no longer cover additional sterol molecules. The shielding would be rendered less effective for DHA-containing PE by the increased molecular area that is associated with the high disorder of the polyunsaturated chain. The poor affinity for cholesterol that the reduced solubility in 16:0-22:6PE implies may also be interpreted in terms of another model suggested by McConnell and coworkers.[26] The model has cholesterol and phospholipid forming complexes that can separate into a complex-rich phase. Because the high disorder of DHA prevents close approach from sterol molecules, the formation of such complexes would be compromised in the case of polyunsaturated phospholipids.

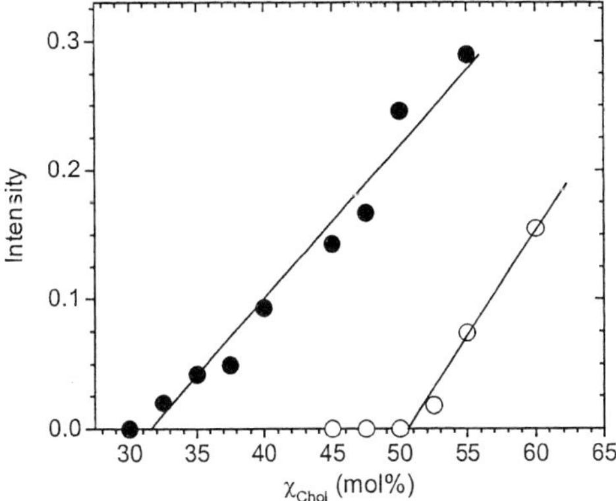

Figure 4. Combined integrated intensity of the second-order XRD peaks 002, 020 and 200 from cholesterol monhydrate excluded from the membrane for 50 wt% aqueous dispersions in 50 mM Tris (pH 7.5) of 16:0-22:6PE/cholesterol at 7.5 °C (●) and 16:0-18:1PE/cholesterol (○) at 40 °C.

DHA Promotes Phase Separation from Lipid Raft Molecules

To support our hypothesis that the introduction of DHA into a membrane enhances the segregation of cholesterol into lipid rafts, we compare the interaction of 16:0-22:6PE vs. 16:0-18:1PE with the sterol in mixtures containing egg sphingomyelin (SM). Figure 5 shows smoothed profiles of order parameter against carbon position for [$^2H_{31}$]16:0-22:6PE/SM (1:1 mol) and [$^2H_{31}$]16:0-18:1PE/SM (1:1 mol) in the absence and presence of cholesterol (1:1:1 mol) at 40 °C. At this temperature both DHA and oleic acid-containing systems are in the lamellar liquid crystalline phase, since SM stabilizes bilayer structure.[27] The order parameter profiles were generated from depaked spectra (inset) on the basis of integrated intensity assuming monotonic dependence.[28] The FFT depaking procedure deconvolutes powder pattern signals to spectra representative of a planar membrane of single alignment.[29] They consist of well-resolved doublets with splittings $\Delta\nu(\theta)$ that equate to order parameters S_{CD} via

$$\Delta\upsilon(\theta) = \frac{3}{2}\left(\frac{e^2qQ}{h}\right)\left|S_{CD}\right|P_2(\cos\theta) \qquad (2)$$

where $\theta = 0^\circ$ is the angle the membrane normal makes with respect to the magnetic field and $P_2(\cos\theta)$ is the second order Legendre polynomial.

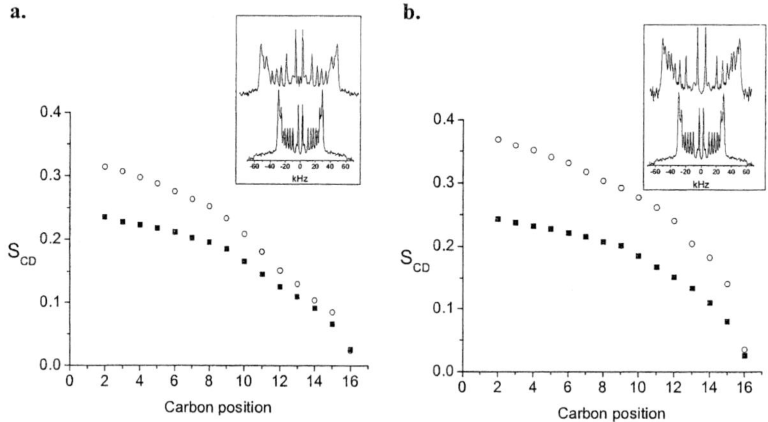

Figure 5. Smoothed order parameter profiles at 40 °C for 50 wt% aqueous dispersions in 50 mM Tris (pH 7.5) of a. [$^2H_{31}$]16:0-22:6PE/SM (1:1 mol) (■) and [$^2H_{31}$]16:0-22:6PE/SM/cholesterol (1:1:1 mol) (○), and b. [$^2H_{31}$]16:0-18:1PE/SM (1:1 mol) (■) and [$^2H_{31}$]16:0-18:1PE/SM/cholesterol (1:1:1 mol) (○). Insets show the depaked spectra (lower and upper without and with cholesterol, respectively) from which the profiles are constructed.

All of the profiles plotted in Figure 5 possess the same characteristic shape. A plateau region of virtually constant order in the top part of the chain is followed by progressively lower order in the bottom part. The profiles demonstrate that cholesterol orders the

[$^2H_{31}$]16:0 sn-1 chain of the PE component in both mixed membranes, but it is clear that the change is markedly less throughout the entire chain for [$^2H_{31}$]16:0-22:6PE (Fig. 5a) than [$^2H_{31}$]16:0-18:1PE (Fig. 5b). The differential corresponds to respective increases of 26% and. 61% in average order parameter. We deduce that the DHA-containing PE separates from the lipid raft molecules SM and cholesterol much more than the oleic acid-containing PE.

Additional evidence is garnered from detergent extraction experiments. This approach identifies sphingolipid/cholesterol-enriched lipid rafts by their insolubility in cold non-ionic detergents such as Triton X-100.[30] Phase separation between detergent resistant membranes (DRM), designated raft-rich, and detergent soluble membranes (DSM), designated non-raft, in 16:0-22:6PE/SM/cholesterol (1:1:1 mol) and 16:0-18:1PE/SM/cholesterol (1:1:1 mol) membranes at 4 °C is assessed in Figure 6. Very little SM or cholesterol (<10%) is found in the DSM fraction for both mixtures, consistent with raft molecules being detergent resistant. A significant difference, on the contrary, exists in the amount of 16:0-22:6PE (70%) vs. 16:0-18:1PE (22%) in the DSM fraction. Greater phase separation from rafts of the DHA-containing phospholipid is again the implication.

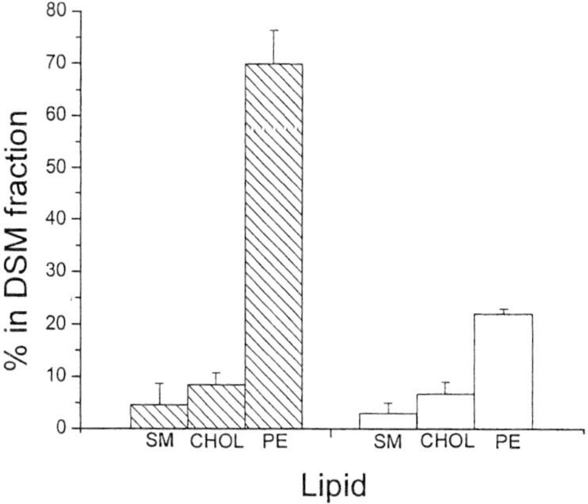

Figure 6. Percentage (mean + standard deviation from three separate experiments) of SM, cholesterol (CHOL) and PE in the DSM fraction of Triton X-100 treated aqueous multilamellar dispersions of 16:0-22:6PE/SM/cholesterol (1:1:1 mol) (shaded) and 16:0-18:1PE/SM/cholesterol (1:1:1 mol) (no pattern) at 4 °C.

We postulate that changes in the lateral architecture of plasma membranes that are driven by the steric incompatibility of PUFA for cholesterol constitute a possible mechanism by which DHA from the diet is able to alleviate a multitude of diseases. A cartoon depicting enhanced segregation of cholesterol into a SM-rich/sterol-rich raft and formation of a DHA-rich/sterol-poor microdomain following the introduction of a DHA-containing phospholipid is drawn in Figure 7. The alteration in local environment has the potential, we propose, to cause signaling proteins to move from a raft to a non-raft region (or vice versa) and consequently undergo conformational modification that modulates cellular signaling events. DHA is known to displace phospholipase D1, for instance, from rafts.[31] The protein is activated in non-raft domains, which disrupts signal transduction and may be the origin of immuno-suppressive effects attributed to DHA.

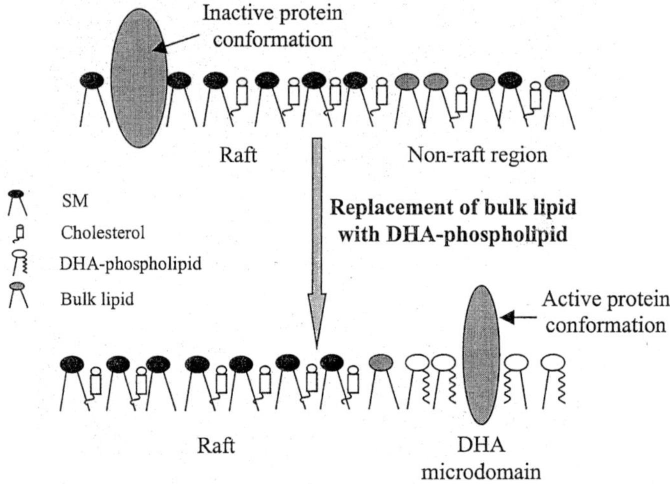

Figure 7. Cartoon rendition of how the introduction of DHA into the outer leaflet of the plasma membrane is proposed to promote the exclusion of cholesterol into SM-rich/sterol-rich rafts and form DHA-rich/sterol-poor microdomains. Concomitant changes in protein location and conformation may have implications for cellular signaling.

Conclusion

The diminished solubility of cholesterol in 16:0-22:6PE (Fig. 3 and 4) and the modest impact upon the phase behavior of [$^2H_{31}$]16:0-22:6PE that equimolar cholesterol elicits (Fig. 1 and 2) unequivocally establish that a highly disordered DHA sn-2 chain deters close contact with the sterol. An appreciably greater tendency for polyunsaturated 16:0-22:6PE than monounsaturated16:0-18:1PE to phase separate from lipid raft molecules SM and cholesterol is furthermore indicated. The sterol-associated increase in order of the sn-1

$[^2H_{31}]$16:0 chain of PE in mixtures with SM is much less for $[^2H_{31}]$16:0-22:6PE compared to $[^2H_{31}]$16:0-18:1PE (Fig. 5), and detergent extraction of PE/SM/cholesterol mixtures at 4 °C finds substantially more of the DHA than the oleic acid-containing PE in the non-raft DSM fraction (Fig. 6). We speculate that these findings may in part explain the diverse health benefits associated with dietary DHA and present a model whereby cellular signaling events may be mediated by DHA-induced relocation of signaling proteins between raft and non-raft regions (Fig. 7).

[1] W. Stillwell, S. R. Wassall, *Chem. Phys. Lipids* **2003**, *126*, 1.
[2] S. R. Shaikh, M. R. Brzustowicz, N. Gustafson, W. Stillwell, S. R. Wassall, *Biochemistry* **2002**, 10593.
[3] S. R. Shaikh, V. Cherezov, M. Caffrey, W. Stillwell, S. R. Wassall, *Biochemistry* **2003**, 12028.
[4] S. E. Feller, K. Gawrisch, A. D. MacKerrell Jr., *J. Amer. Chem. Soc.* **2002**, *124*, 318.
[5] T. Huber, K. Rajamoorthi, V. F. Kurze, K. Beyer, M. F. Brown, *J. Amer. Chem. Soc.* **2002**, *124*, 298.
[6] E. A. Dratz, A. J. Deese, in: *"Health Effects of Polyunsaturated Fatty Acids in Seafoods"*, A. P. Simopolous, R. R. Kifer, R. E. Martin, Eds., Academic Press, New York 1986, p. 319.
[7] A. Salmon, S. W. Dodd, G. D. Williams, J. M. Beach, M. F. Brown, *J. Amer. Chem. Soc.* **1987**, *109*, 2600.
[8] L. L. Holte, S. A. Peter, T. M. Sinnwell, K. Gawrisch, *Biophys. J.* **1995**, *68*, 2396.
[9] D. Huster, J. J. Albert, K. Arnold, K., K. Gawrisch, *Biophys. J.* **1997**, 73, 856.
[10] V. T. Armstrong, M. R. Brzustowicz, S. R., Wassall, L. J, Jenski, W. Stillwell, *Arch. Biochem. Biophys.* **2003**, *414*, 74.
[11] B. W. Koenig, H. H. Strey, K. Gawrisch, *Biophys. J.* **1997**, *73*, 1954.
[12] M. R. Brzustowicz, W. Stillwell, S. R. Wassall, *FEBS Lett.* **1999**, *451*, 197.
[13] M. R. Brzustowicz, V. Cherezov, M. Caffrey, W. Stillwell, S.R. Wassall, *Biophys. J.* **2002**, *82*, 285.
[14] M. R. Brzustowicz, V. Cherezov, M. Zerouga, M. Caffrey, W. Stillwell, S. R. Wassall, *Biochemistry* **2002**, *41*, 12509.
[15] S-L. Niu, B. J. Litman, *Biophys. J.* **2002**, *83*, 3406.
[16] K. Simons, W. L. C. Vaz, *Annu. Rev. Biophys. Biomol. Struct.* **2004**, *33*, 269.
[17] D. Huster, K. Arnold, K. Gawrisch, *Biochemistry* **1998**, *37*, 17299.
[18] D. C. Mitchell, B. J. Litman, *Biophys. J.* **1998**, *75*, 896.
[19] P. McLennan, T. Bridle, M. Abeywarden, J. Charnock, *Amer. J. Clin. Nutr.* **1993**, *58*, 666.
[20] J. H. Davis, *Biochim. Biophys. Acta* **1983**, *737*, 117.
[21] R. L. Thurmond, G. Lindblom, M. F. Brown, *Biochemistry* **1993**, *32*, 5394.
[22] T. P. W. McMullen, R. N. McElhaney, *Curr. Opin. Colloid Interface Sci.* **1996**, *1*, 83.
[23] B. M. Craven, *Nature* **1976**, *260*, 727.
[24] J. Huang, J. T. Buboltz, G. W. Feigenson, *Biochim. Biophys. Acta* **1999**, *1417*, 89.
[25] J. Huang, G. W. Feigenson, *Biophys. J.* **1999**, *76*, 2142.
[26] H. M. McConnell, M. Vrljic, *Annu. Rev. Biophys. Biomol. Struct.* **2003**, *32*, 469.
[27] S. R. Shaikh, A. C. Dumaual, A. Castillo, D. LoCascio, R. Siddiqui, W. Stillwell, S. R. Wassall, *Biophys. J.* **2004**, *87*, 1752.
[28] M. Lafleur, B. Fine, E. Sternin, P. R. Cullis, M. Bloom, *Eur. Biophys. J.* **1989**, 56, 1037.
[29] M. A. McCabe, S. R. Wassall, *Solid State Nucl. Magn. Reson.* **1997**, *10*, 53.
[30] S. N. Ahmed, D. A. Brown, E. London, *Biochemistry* **1997**, *36*, 10944.
[31] O. Diaz, A. Berquand, M. Dubois, S. Di Agostino, C. Sette, S. Bourgoin, M Lagarde, G. Nemoz, A-F. Prigent, *J. Biol. Chem..* **2002**, *277*, 39368.

Domain Size and Fluctuations at Domain Interfaces in Lipid Mixtures

*Heiko M. Seeger, Matthias Fidorra, Thomas Heimburg**

Niels Bohr Institute, University of Copenhagen, Blegdamsvej 17, 2100 Copenhagen Ø, Denmark, and Max-Planck Institute for Biophysical Chemistry, 37077 Göttingen, Germany
E-mail: theimbu@nbi.dk

Summary: Biomembranes consist of a complex mixture of a large number of lipids and proteins. In such mixtures, microscopic domains and macroscopically separated phases may exist. Here, we discuss phase behavior and domains formation of binary lipid mixtures. We show that the domain formation is accompanied by large fluctuations at the domain boundaries, resulting in altered physical properties at the boundaries, for instance in a pronounced increase of the elastic constants. Therefore, we argue that the physics of the membrane depends on the overall length scale of its domains interfaces. We present here confocal microscopy images, calorimetric melting profiles and Monte-Carlo simulations to understand the factors that determine domain formation, their sizes and the role of the domain interfaces.

Keywords: confocal microscopy; domains; elastic constants; fluctuations; phase diagrams

Abbreviations: DLPC - dilaureoyl phosphatidylcholine; DMPC – dimyristoyl phosphatidylcholine; DPPC - dipalmitoyl phosphatidylcholine; DSPC – distearoyl phosphatidylcholine

Introduction

Biomembranes consist of a large variety of different lipids and proteins. The composition of each membrane is distinctively different, even between membranes of the various organelles within one cell [1]. About 80% of the lipids in eukaryotic membranes are zwitterionic phosphatidylcholines or phosphatidylethanolamines, the rest is uncharged or charged. Mitochondrial membranes, for instance, are rich in charged lipids (about one net negative charge per five lipid chains), whereas plasma membranes have typically less than one net negative charge per twenty lipid chains. Plasma membranes, on the other hand, are usually rich in cholesterol (up to 30% of the lipids), and there exist differences between inner and outer leaflets of the bilayers. The reason for the large diversity in lipid

 DOI: 10.1002/masy.200550109

composition is subject to an ongoing debate. In recent years it became increasingly evident that the lipid heterogeneity gives rise to domain formation in the membrane plane, which may influence diffusion pathways and communication networks between proteins and other molecules. More and more experimental data show the existence of domains in biomembranes, and that they are in fact important for membrane function. The finding of phase separation in lipid membranes is not at all new [2]. Domains were already visualized in monolayers in the 1980s [3]. First pictures of domains in vesicles, however, were published only 1999 in pioneering works by Korlach et al. [4], and by Bagatolli and Gratton [5]. In artificial lipid mixtures, domains are usually much larger and easier to observe with confocal microscopy than in intact biomembranes. In the biology community domains in biomembranes are frequently called “rafts”. Their sizes are typically below microscopic resolution and evidence for their existence is rather indirect. Many of those micro-domains seem to be rich in cholesterol, sphingolipids and special proteins [6]. However, it is not at all obvious that domains in biomembranes by necessity display a unique composition, and the finding of such specific domains may be artifacts from the preparation method, namely detergent extraction [7]. The term “raft” furthermore implies that they are often considered as stable functional units, comparable to large protein complexes.

The question arises whether the fact that domains in biomembranes are small is not in conflict with the assumption that they are rigid objects. In fact, in lipid mixtures fluctuations may be very large and the magnitude of fluctuations relates to the size of the fluctuating objects (relative fluctuations become larger for smaller objects). In some simple lipid systems macroscopic phase separation has been observed [8], meaning that the domain size is on the order of the vesicle size and the membrane demixes into just two macroscopic regions with distinctly different physical properties or order parameters. In such systems the domain size grows with vesicle size. In the thermodynamic limit, when a vesicle is infinitely large, the length and the physical properties of the domain interface can be neglected in relation to the overall features of the domains. Such macroscopic domains are called phases. The number of coexisting phases depends on the number of components and the number of degrees of freedom. The Gibbs phase rule applied to a system observed at constant pressure is given by

$$F = K - P + 1 \quad ,$$

where F is the number of degrees of freedom, K is the number of components (=2 in a binary lipid system in excess water) and P is the number of coexisting phases. In a binary system P may be 1,2 or 3 (at eutectic or peritectic points). One complication in using Gibbs' phase rule is that a membrane may exhibit changes in curvature [8] and thus may display one more degree of freedom. Gibbs' phase rule was also derived neglecting phase boundaries, which can only be done if phase separation is macroscopic. Assuming macroscopic phase separation, one can use phase diagrams to calculate the relative fractions of each phase and their composition, making use of thermodynamic constructions, e.g. the lever rule [2]. The thermodynamic treatment becomes more complicated when domains are small and do not scale with system size. In the melting of a one-component membrane this case corresponds to a continuous transition (domains in the melting regime are smaller than system size and there is no latent heat) in contrast to a first order transition with macroscopic domain formation and latent heat [9]. In Monte-Carlo simulations one can distinguish first order melting from continuous transitions by analyzing histograms of the distribution of vesicular states. Continuous transitions display Gaussian fluctuations around the thermal equilibrium at all temperatures, whereas first order transitions display coexisting macroscopic states at the transition temperature [9]. In systems containing several components there also exists a fundamental difference between cases where domain sizes grow with system size as compared to cases where they are independent of scaling. In the second case, the properties of the domain interfaces can never be neglected and Gibbs' phase rule cannot adequately be applied.

This paper focuses on the factors that determine domain sizes, and where the differences between microscopic and macroscopic domain formation arise. To this purpose we compare heat capacity profiles with Monte-Carlo simulations and with Confocal Microscopy images that demonstrate domain formation in giant vesicles. In particular, we focus on the fluctuations at domain interfaces.

Materials and Methods

Calorimetric experiments were performed using a Hart Scientific (Provo, Utah) four cell scanning calorimeter and a VP differential scanning calorimeter by MicroCal (Northampton, MA) using scan rates of 5 deg/hr. The lipid mixtures for calorimetry were dried from organic solvent (dichloromethane-methanol 2:1 mixtures). Samples were measured in distilled water at neutral pH or in a 10mM Hepes, 1mM EDTA buffer at pH 7.4. Vesicles for confocal microscopy were prepared on Indium Tin Oxide (ITO) cover slips with the electroformation method [10]. The lipid mixtures for confocal microscopy (including a fraction of 10^{-4} of fluorescence dyes specific for gel and fluid domains) were dried on the cover slip from organic solvent (dichloromethane-methanol 1:1 mixtures). Remaining solvent was removed in a high vacuum desiccator. The dry lipid films were hydrated and exposed to an alternating electrical field of 10 Hz and 3V [10]. Images were recorded using confocal microscopes by Leika and Zeiss.

The lipid melting behavior was modeled using Monte-Carlo simulations. Here, we employ an Ising-like two-state model, where each chain may be either in a gel state (ordered chains) or a fluid state (disordered chains). The two chain states are different in enthalpy and entropy. Furthermore, nearest neighbor interactions are defined that lead to interaction parameters, which are responsible for cooperative melting. In a two-component membrane, there are two states for each species, and 6 unlike nearest neighbor interactions. The Gibbs free energy of each configuration of such a system is given by:

$$\begin{aligned} G = n_A^f \cdot (\Delta H_A - T\Delta S_A) + n_B^f \cdot (\Delta H_B - T\Delta S_B) + \\ + n_{AA}^{gf}\omega_{AA}^{gf} + n_{AB}^{gg}\omega_{AB}^{gg} + n_{AB}^{gf}\omega_{AB}^{gf} + n_{AB}^{fg}\omega_{AB}^{fg} + n_{AB}^{ff}\omega_{AB}^{ff} + n_{BB}^{gf}\omega_{BB}^{gf} \end{aligned} \tag{1}$$

where n_A^f and n_B^f are the numbers of lipid chains of species A and B in a fluid state. The $n_{\alpha\beta}^{ij}$ are the number of interactions between species α and β in state i and j, respectively. Four additional parameters are the melting enthalpies of the two species, ΔH_A and ΔH_B, and their melting entropies, ΔS_A and ΔS_B. These parameters can all be determined from the experimental melting profiles. In the Monte-Carlo procedure, a given number of lipids of species A and B is distributed randomly on a computer generated triangular lattice with periodic boundary conditions. Then, attempts are made to change the state of a randomly chosen lipid from fluid to gel or vice versa (Glauber steps). The likelihood of such an event to happen is given by a Boltzmann factor. Furthermore, diffusion is modeled by nearest neighbor exchange (Kawasaki steps). Details are given in [11,12,13].

Theoretical considerations

During a Monte-Carlo simulation the membrane system fluctuates around the thermal equilibrium, e.g. each snapshot during the simulation displays a slightly different enthalpy. The fluctuations in enthalpy, however, are proportional to the heat capacity, c_P, via the relation [14]

$$c_P = \frac{\overline{H^2} - \overline{H}^2}{RT^2} \tag{2}$$

Similarly, fluctuations in volume yield the isothermal volume compressibility, κ_T^V, and fluctuations in area yield the isothermal area compressibility, κ_T^A [14]:

$$\kappa_T^V = \frac{\overline{V^2} - \overline{V}^2}{\overline{V} \cdot RT} \quad \text{and} \quad \kappa_T^A = \frac{\overline{A^2} - \overline{A}^2}{\overline{A} \cdot RT} \; . \tag{3}$$

These relations are a consequence of the fluctuation-dissipation theorem [15,16], but can also be derived just from the derivatives of the statistical thermodynamics averages of H_i, V_i and A_i (i denoting the different states of the system). According to Evans [17], the bending elasticity, κ_B (or the bending modulus, $K_B = 1/\kappa_B$) of a membrane is related to the isothermal area compressibility via

$$\frac{1}{K_b} = \kappa_b = \frac{16 \cdot \kappa_T^A}{D^2} \quad , \tag{4}$$

where D is the membrane thickness (cf. [14]).

It has been shown experimentally that excess enthalpy changes, ΔH, excess volume changes, ΔV, and excess area changes, ΔA during melting transitions are (within experimental error) exact proportional functions of the temperature ($\Delta V(T) = \gamma_V \cdot \Delta H(T)$, $\Delta A(T) = \gamma_A \cdot \Delta H(T)$). Therefore, enthalpy fluctuations, volume and area fluctuations are also proportional functions of the temperature.

Therefore, we obtain the following useful relations for changes of the elastic constants in melting transitions:

$$\Delta \kappa_T^V = \frac{\gamma_V^2 \; T}{V} \cdot \Delta c_P \quad , \tag{5}$$

$$\Delta \kappa_T^A = \frac{\gamma_A^2 \; T}{A} \cdot \Delta c_P \quad , \tag{6}$$

$$\Delta \kappa_B = \frac{16 \; \gamma_A^2 \; T}{D^2 A} \cdot \Delta c_P \; , \tag{7}$$

where $\gamma_V = 7.8 \cdot 10^{-4} cm^3 / J$ and $\gamma_A = 8.9 \cdot 10^3 cm^2 / J$ are constants, roughly independent on the choice of the lipid [14,18].

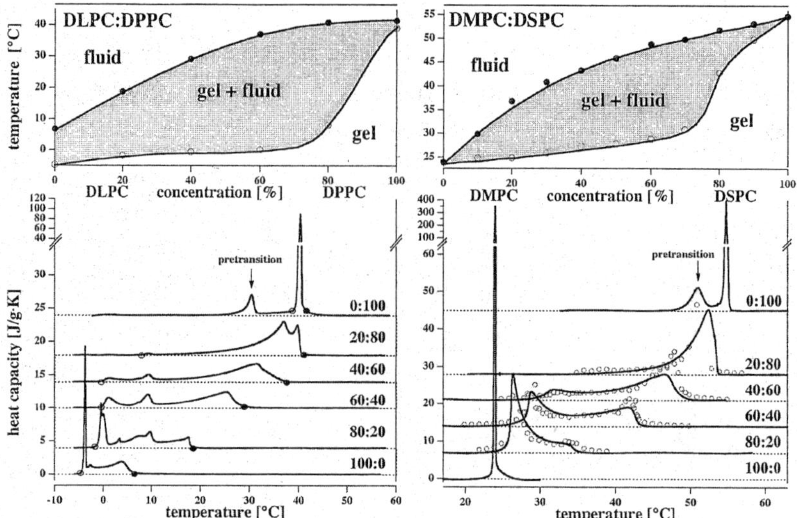

Figure 1: Left: Phase diagrams of DLPC-DPPC mixtures and the corresponding heat capacity profiles. The upper and lower phase boundaries have been obtained from the upper (●) and lower (O) limits of the c_P-profiles. **Right:** Phase diagram of DMPC-DSPC mixtures and the corresponding heat capacity profiles. Symbols (O) represent results from Monte-Carlo simulations. The phase coexistence regime is shaded in grey.

The fluctuations given in eqs. (2) and (3) are those of the whole lipid matrix, thus being macroscopic quantities. However, one can also define the fluctuations at each lattice point of the simulation by defining a state parameter S(i,j) for each lattice site, which may be either 0 or 1 for gel and fluid lipid state, respectively [13]. The degree of fluctuations at each lattice point can be determined during a Monte-Carlo simulation by determination of the mean square deviation of the parameter S(i,j).

$$\text{local fluctuations} = \overline{S^2(i,j)} - \overline{S(i,j)}^2 \qquad (8)$$

Fluctuations as defined in (8) may assume values between 0 and 0.25. Large local fluctuations are accompanied by increased elastic constants and reduced relaxation times [19]. Using Monte-Carlo simulations we show below that such local fluctuations are especially strong at domain interfaces, thus giving rise to distinctly different physical properties at the domain boundaries. Practically, in the simulation the local fluctuations were recorded such that diffusion steps were switched off after equilibration for some MC-cycles and the state fluctuations in a given lipid configuration were determined before switching diffusion on again.

Results

Phase behavior of binary lipid mixtures

In the following we will consider the phase behavior of binary lipid systems, in particular of DLPC:DPPC and DMPC:DSPC mixtures. We recorded heat capacity profiles at various molar ratios (Fig.1).

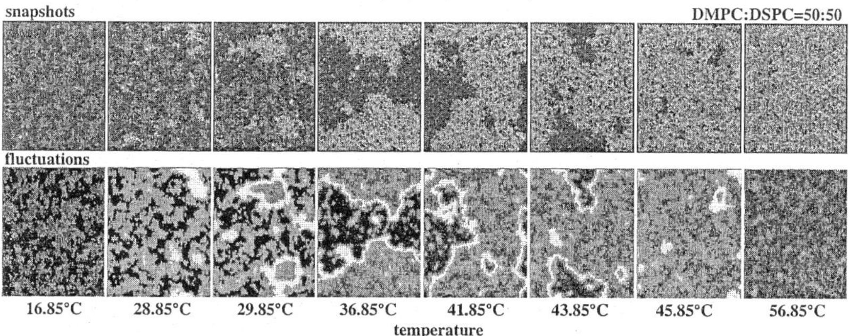

Figure 2: Lipid state distribution and fluctuations of a DMPC:DSPC=50:50 mixture in the chain melting regime as a function of temperature (from Monte-Carlo simulations). **Top:** Snapshots at different temperatures with a simulation box of 80×80 lipid chains. The two dark grey shades correspond to gel state lipid chains of DMPC and DSPC, respectively. The two light grey shades correspond to the fluid states of DMPC and DSPC. **Bottom:** Fluctuations of the lipid matrices in the top panels, as defined in eq. 8. Brighter shades correspond to larger fluctuations. Fluctuations are largely enhanced at the domain interfaces.

A common procedure to generate phase diagrams is to identify phase boundaries from a tangent construction at the lower and upper limits of the heat capacity profiles. These values are plotted into a temperature versus concentration graph. This is an empirical method that does not guarantee that these temperature limits are the correct phase boundaries. If phase separation occurs, the composition of the macroscopic phases is well defined. Under such a condition, from confocal microscopy determination of domain sizes as a function of temperature and composition one can draw phase diagrams that relate the position in phase space with fraction of each phase and its composition, and compare it to calorimetric measurements.

Heat capacity profiles can also be simulated with reasonable accuracy using Monte-Carlo simulations as described above (see [11,12,13], cf. Fig. 1, right). Snapshots produced in such simulations (Fig. 2) can be used to arrive at a deeper understanding of the processes that result in domain formation, and they can be compared to confocal microscopy images (Fig. 3). Fig. 2 (top row) shows how domains form as a function of temperature in the DMPC:DSPC=50:50 mixture. At temperatures close to the outer heat capacity maxima (~29°C and ~44°C, respectively, cf. Fig. 4), gel and fluid domains are smaller than system size. In the intermediate temperature regime domains become macroscopic (only one large gel and fluid domain – note the periodic boundary conditions).

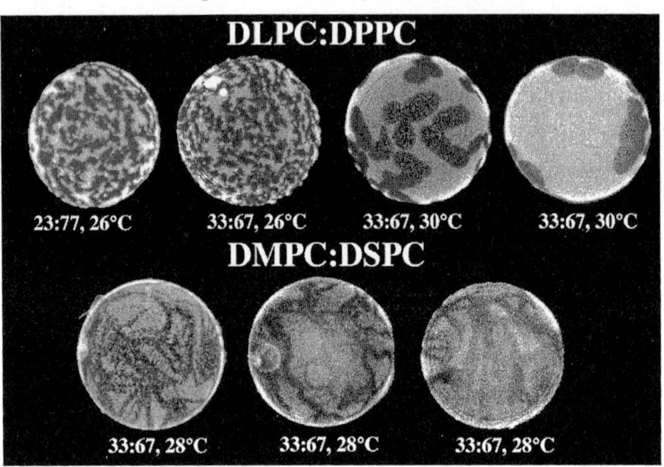

Figure 3: Confocal microscopy images of giant vesicles of DLPC-DPPC and DMPC-DSPC mixtures at different mixing ratios and temperatures. Domain sizes and shapes vary considerably, dependent on experimental conditions. Dark regions correspond to gel domains. Vesicular sizes range from 20 to 30 μm.

Within each phase, however, one can still notice small domains of the opposite chain state, which fluctuate during simulation time. Even in the gel phase at low (16.85°C) and the fluid phase at high (56.85°C) temperatures one can recognize local segregation of the two lipids into two types of gel or fluid domains, respectively. The bottom row of Fig. 2 shows the local fluctuations (eq. 8) corresponding to the snapshots in the top row. It can be clearly seen that they occur predominantly at the domain interfaces. Remember that highly fluctuating regions will display the highest elasticity and compressibility. The simulations shown here do not contain out-of-plain degrees of freedom. However, in coupled bilayers curvature modes can in principle be obtained by considering the local area differences on both monolayers. This concept has successfully been used by Heimburg [20] to obtain a consistent thermodynamical picture of the formation of the ripple phase. By looking at the temperature progression of the fluctuations (Fig. 2) one can well see that domain formation is not the only property that describes the physics of the system but that there are also large fluctuations in particular in temperature regimes of high heat capacity, which not necessarily coincide with regions of large domain formation. Those are the most interesting regimes.

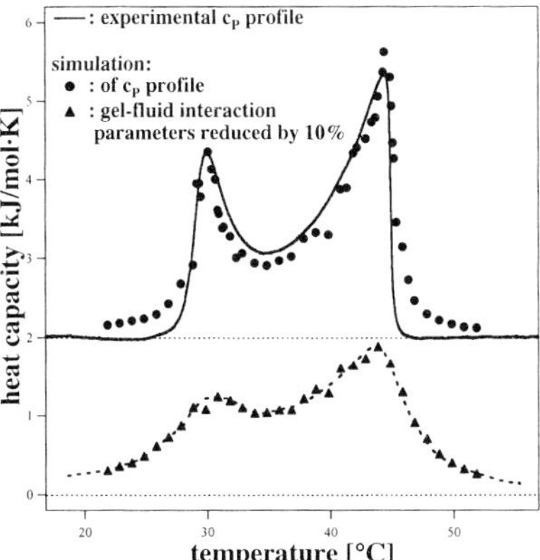

Figure 4: Top: Experimental heat capacity profile of a 50:50 DMPC:DSPC mixture (solid line) and the corresponding Monte-Carlo simulation (symbols). **Bottom:** Simulation of the same mixture using gel-fluid interaction parameters, $\omega_{\alpha\beta}^{gf}$, being only 90% of those used to simulate the top trace.

Confocal microscopy images of lipid mixtures demonstrate that subtle changes in temperature can alter the vesicles from forming domains smaller than vesicular size (DLPC:DPPC=33:67, 26°C) to vesicles displaying large domains (DLPC:DPPC=33:67, 30°C). Gel and fluid domain areas deviate from the values obtained by applying the lever rule to the phase diagram in Fig.1 (left).

In simulations it can be shown that the criteria for whether a system displays microscopic domains or macroscopic domains (that grow with system size) depends in a delicate manner on the interaction between lipids (Fig. 4 and Fig. 5). The size and shape of domains is dominated by nearest neighbor interactions in the lipid matrix. Fig. 4 shows that the melting profiles of a DMPC:DSPC=50:50 (top trace) mixture is slightly broadened upon reduction of the gel-fluid interactions by 10% (bottom trace) in the simulation. The consequences for the domain pattern, however, are dramatic. Fig.5 shows Monte-Carlo snapshots for varying matrix sizes (30×30, 60×60, 120×120 and 180×180 chains) for the two cases in Fig.4 at T=36.85°C. The parameters that describe the experimental profiles well (top trace in Fig. 4) lead to macroscopic phase separation (Fig.5, left top), evident in the separation into one large gel and one large fluid domain, independent of matrix size.

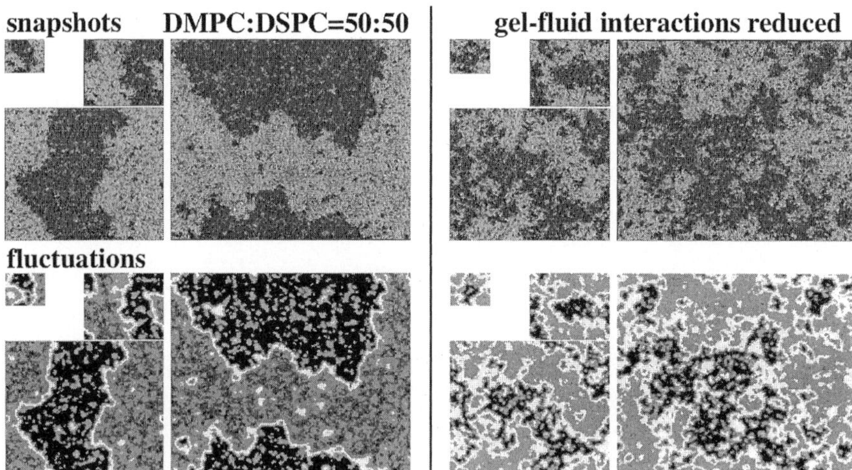

Figure 5: Simulation snapshots for various matrix sizes: 30 × 30, 60 × 60, 120 × 120 and 180 × 180 lipid chains. **Left, top**: Snapshots corresponding to the heat capacity curve of the DMPC:DSPC=50:50 mixture at 36.9°C (upper c_P-profile of Fig. 4), showing macroscopic phase separation into fluid and gel independent of matrix size. **Left, bottom:** Corresponding local fluctuations. **Right, top:** Snapshots corresponding to the bottom c_P trace of Fig.5 at 36.85°C, showing that domains are smaller than the matrix size on all length scales if the interfacial tension at the gel-fluid domain boundaries are reduced. **Right, bottom:** Corresponding fluctuations, showing pronounced differences in the fluctuations as compared to the case in the left hand panels.

The fluctuations in this system are large at the domain interfaces (white regions in Fig.5, left, bottom). Increasing matrix size leads to a relative reduction of the fraction of the fluctuating interface. Interestingly, the two phases themselves display different fluctuation strength (evident from different shades of grey in the bottom panels of Fig.5). If the gel-fluid nearest neighbor interactions are reduced by 10% in the simulation (as it could be done experimentally by adding small molecules like local anesthetics), the differences in domain shape are profound (Fig.5, right). Domain formation now occurs on all length scales and there are no obvious dependencies on matrix size. Although we did not perform correlation length analysis, this can in principle be done by using pair correlation functions [21] or by calculating the structure factor of the lipid matrix [22]. The composition of each domain fluctuates and is not well defined as compared to macroscopic phase separation (Fig.5, left). It is unlikely that confocal microscopy would be able to correctly determine gel and domain areas. One cannot consider these systems as phase separated because on all length scales coexistence of the two lipid states is found. The lipid matrix is to a much larger degree dominated by fluctuating interfaces as compared to the left hand panels of Fig.5. This effect is even more pronounced if the gel-fluid interactions are reduced by 20% (not shown), where the large-scale domain structure nearly completely dissolves and the whole matrix is dominated by fluctuations. As mentioned, such changes can among other factors be induced by addition of small molecules, which broaden c_P-profiles and thus reduce the interfacial tension at domain boundaries (cf. 13).

Conclusions

Here, we present "phase diagrams" of lipid mixtures derived from calorimetry and critically discuss the implications of a statistical thermodynamics analysis of the heat capacity profiles. Clearly, the reduction of domain length scales induced by reduction of the interaction parameters (interfacial tension at the domain boundaries) leads to larger fluctuations of the lipid systems in the melting regime. No defined physical property (e.g. defined concentration of components, or order parameter) can be attributed to the individual domains. Thus, the concept of phase separation starts failing if the domains become small. This is caused by the fact that the lipid matrix starts to be dominated by domain interfaces and their fluctuations, which do not have an equivalent in a picture consisting exclusively of coexisting phases. In respect to the ongoing discussion of the physical nature of "rafts" in biomembranes, it is likely that small domains (or rafts) will be subject to large fluctuations. In a recent review by Simons and Vaz [6] the authors argue *"There is no fundamental requirement that any phase in a heterogeneous system not be divided into several part or domains, although their exact thermodynamic description becomes unreliable when the domains become too small"*. In our view the first part of this statement is incorrect, while the second one is clearly true. This is the case because the fluctuations at the domains interfaces enter into the physical picture.

[1] E. Sackmann, in: Structure and Dynamics of Membranes: From Cells to Vesicles, Elsevier **1995**
[2] A. G. Lee, Biochim.Biophys. Acta **1977**, 472, 285
[3] H. M. McConnell, D. J. Keller, Proc.Natl.Acad.Sci.USA **1987**, 84, 4706
[4] J. Korlach, P. Schwille, W. W. Webb, G. W. Feigenson, Proc.Natl.Acad.Sci.USA **1999**, 96, 8461
[5] L. A. Bagatolli and E. Gratton, Biophys. J. **1999**, 77, 2090
[6] K. Simons, W. L. C. Vaz, Annu.Rev.Biomol.Struct. **2004**, 33, 269
[7] H. Heerklotz, Biophys.J. **2002**, 83, 2693
[8] T. Baumgart1, S. T. Hess, W. W. Webb, Nature **2003**, 425, 821
[9] V. P. Ivanova, T. Heimburg, Phys Rev. E **2001**, 63, 1914
[10] M. I. Angelova, S. Soléau, P. Méléard, J. F. Faucon, P. Bothorel, Progr. Colloid. Polym. Sci. **1992**, 89, 127
[11] I. P. Sugar, T. E. Thompson, R. L. Biltonen. Biophys. J. **1999**, 76, 2099
[12] A. E. Hac, H. Seeger, M. Fidorra, T. Heimburg. Biophys. J. **2004**, submitted
[13] V. P. Ivanova, I. Makarov, T. E. Schäffer, T. Heimburg, Biophys. J. **2003**, 84, 2427
[14] T. Heimburg, Biochim. Biophys. Acta **1998**, 1415, 147
[15] H. B. Callen, R. F. Green, Phys. Rev. **1952**, 86, 702
[16] R. Kubo, Rep. Prog. Phys. Phys. **1966**, 29, 255
[17] E. A. Evans, Biophys. J. **1974**, 14, 923
[18] H. Ebel, P. Grabitz, T. Heimburg. J. Phys. Chem. B **2001**, 105, 7353
[19] P. Grabitz, V. P. Ivanova, T. Heimburg, Biophys. J. **2002**, 82, 299
[20] T. Heimburg, Biophys. J. **2000**, 78, 1154
[21] K. Jørgensen, M. M. Sperotto, O. G. Mouritsen, J. H. Ipsen, M. J. Zuckermann, Biochim. Biophys. Acta **1993**, 1152, 135
[22] L. K. Nielsen, T. Bjørnholm, O. G. Mouritsen, Nature **2000**, 404, 352

Probing the Interdigitated Phase of a DPPC Lipid Bilayer by Micropipette Aspiration

*Hung V. Ly, Marjorie L. Longo**

Department of Chemical Engineering and Material Science, University of California, Davis, CA 95616, USA
E-mail: mllongo@ucdavis.edu

Summary: We used micropipette aspiration of giant unilamellar vesicles to directly measure the areal expansion of gel ($L_{\beta'}$) phase 1,2-dipalmitoyl-sn-glycero-3-phosphocholine (DPPC) lipid bilayers induced by exposure to ethanol/water mixtures. Areal expansion began in 7 vol% ethanol and increased monotonically as the concentration of ethanol was increased to 15 vol% at which point areal expansion reached a plateau of 50%. This ethanol concentration range is in good agreement with that of the interdigitated phase ($L_{\beta}I$) of DPPC, therefore, we believe that this is the first direct measurement of the areal expansion accompanying interdigitation of gel-phase lipids. Our observations are consistent with the presence of coexisting $L_{\beta}I$ and $L_{\beta'}$ phases in ethanol concentrations between 7% and 15 vol% and 100% $L_{\beta}I$ phase in 15 vol% ethanol and higher. We observed a bimodal distribution of areal expansion (0% and 20%) induced by 7 vol% ethanol indicating that at the threshold concentration, interdigitation is induced in only a portion of DPPC vesicles. Areal expansion could not be easily reversed, consistent with kinetic trapping of the $L_{\beta}I$ phase. DPPC vesicles exposed to butanol at the known threshold and plateau concentrations for the $L_{\beta}I$ phase displayed areal expansion behavior consistent with our ethanol observations. However, the area expanded significantly faster for DPPC bilayers exposed to butanol vs. ethanol, which we attribute to enhanced partitioning of the longer-chained butanol into the lipid headgroups. Ethanol-induced areal expansion of DPPC bilayers was inhibited by inclusion of 10 mol% and 25 mol% cholesterol in the bilayer. However, areal expansion could be induced by application of tensions (~8 mN/m) similar to the phenomena of interdigitation induced by high pressure. The presence of 20 vol% ethanol significantly decreased surface cohesion of DPPC bilayers containing 25 mol% cholesterol as evidenced by a decreased area compressibility modulus and lysis tension.

Keywords: area compressibility modulus; cholesterol, DPPC; ethanol; phase separation; rupture tension; Traube's rule

Introduction

Phosphatidylcholine (PC) lipids can exist in a phase in which the acyl chains from the opposing halves of the bilayer are fully interdigitated.[1] A transition from the gel-phase ($L_{\beta'}$) to the interdigitated phase ($L_{\beta}I$) is induced by the addition of short-chain alcohols or

 DOI: 10.1002/masy.200550110

other small molecules (ethylene glycol, Tris, anesthetics, etc.) to the surrounding aqueous media, or high hydrostatic pressure.[2] The physical structure and thermodynamic properties of the $L_{\beta}I$ phase have been well characterized by many techniques, including X-ray diffraction,[1] neutron diffraction,[3] fluorescence probes,[4] and differential scanning calorimetry (DSC).[5] Many of these studies were aimed at determining the degree of membrane thinning and phase transition temperature as well as the threshold concentrations and pressure of the $L_{\beta'}$-$L_{\beta}I$ phase transition. The mechanism behind the induction of the $L_{\beta}I$ phase has been attributed to the binding of the short-chain alcohols to the phospholipid bilayers in the $L_{\beta'}$ phase. The presence of the alcohols in the headgroup region increases the lateral and transverse repulsion between the lipid molecules that is first relieved by the tilting of the hydrocarbon chain and followed by chain interdigitation.[6,7] Biologically, the substantial thickness changes that accompany $L_{\beta'}$-$L_{\beta}I$ transition (e.g. 33%) should strongly affect the normal functions of membrane-associated proteins due to the mismatch of the hydrophobic region. Therefore, the role of the $L_{\beta}I$ phase in alcohol toxicity, alcohol tolerance, general anesthesia, cell viability, food sterilization, membrane fusion, and metabolic changes is of great interest.[8-10] In addition, the $L_{\beta'}$-$L_{\beta}I$ phase transition has been utilized in the formation of drug delivery vehicles.[11] It has also been found that the transition can be inhibited by the addition of cholesterol (Chol) in PC bilayers where the cholesterol alleviates the lipid head-group crowding that favors interdigitation.[12,13]

Inspired by our previous work where we demonstrated that short-chain alcohols (methanol, ethanol, propanol, and butanol) can significantly modify the mechanical properties and increase the area per molecule of fluid-phase PCs,[14] we now explore the $L_{\beta'}$-$L_{\beta}I$ phase transition and $L_{\beta}I$ phase of gel-phase PCs by micropipette aspiration (MPA). For a general review and the theoretical framework of the technique, see Needham and Zhelev.[15,16] Unlike other techniques (e.g. X-ray diffraction or neutron diffraction) that average over a large population of multilamellar vesicles, the MPA technique provides a unique opportunity to probe directly the structural and mechanical properties of the single bilayer of individual giant vesicles from 20 to 40 μm in diameter. MPA has been applied successfully to measure the mechanical and viscous properties of natural and synthetic systems (i.e. red blood cells,[17] egg lecithin vesicle,[18] and diblock polymers[19]) and the adsorption of small molecules into the bilayer and the accompanying area expansion (e.g.

the influenza hemagglutinin fusion peptide[20-22] and lysolipid[23]). Using MPA on giant unilamellar DPPC vesicles, we observe substantial area expansion (20-80%) of DPPC membranes undergoing the $L_{\beta'}$-$L_{\beta}I$ phase transition when exposed to ethanol-water or butanol-water mixtures and subsequent kinetic trapping of the $L_{\beta}I$ phase. We find that the $L_{\beta'}$-$L_{\beta}I$ phase transition is not induced in all vesicles at the threshold concentration of alcohol. In addition, the rate of membrane area expansion is substantially faster for exposure to butanol than exposure to ethanol. We explore the inhibiting effect of cholesterol (10 and 25 mol%) and show that it is possible for ethanol to decrease the mechanical cohesion of DPPC membranes containing cholesterol. It is shown that the inhibitory effect of cholesterol can be overcome by application of membrane tension. Also discussed are morphologies of DPPC vesicles in aqueous and ethanol-water environments and the lysis tensions and area compressibility moduli of DPPC, DPPC/10 mol% Chol, and DPPC/25 mol% Chol.

Experimental Section

Materials

1,2-dipalmitoyl-sn-glycero-3-phosphocholine and cholesterol in chloroform were purchased and used without further purification from Avanti Polar Lipids (Alabaster, AL). Ethanol (200 Proof) was bought from Gold Shield Chemical Co. (Hayward, CA). 1-butanol, sucrose, and glucose of high grade were bought from Sigma-Aldrich (St. Louis, MO). Chloroform and methanol were purchased from Fisher Scientific (Fairlawn, NJ). Bovine serum albumin (fraction V, low heavy metals) was purchased from Calbiochem (San Diego, CA). Deionized water used was produced by a Barnstead nanopure water system (Dubuque, IA) and had a resistivity of 18.1 MΩ-cm.

Giant Vesicles Preparation

We formed giant unilamellar vesicles composed of DPPC and Chol by the electroformation technique.[24] Briefly, stock solutions of DPPC and Chol were both diluted to 0.5 mg/ml solution in chloroform/methanol (2:1 volume ratio). Appropriate amounts of DPPC and Chol were premixed to have solutions with 9:1 and 3:1 (mol:mol) DPPC:Chol. We spread 50 μl of the premixed solution onto two platinum wires housed in an open Telfon® block that was then placed under vacuum for over 2 hours to remove any trace amount of solvent. We sealed the open volume around the wires in the block with coverslips coated with SurfaSil® (Pierce, Rockford, IL). We next filled the volume with 100 mM sucrose aqueous solution (the vesicle's interior solution) that was preheated to ~50°C, above DPPC's phase transition temperature of 41°C.[25] Afterward, we submerged the block in a glass bowl filled with 100 mM sucrose aqueous solution, preheated to ~50°C. With the glass bowl in an oven set at ~50°C, we applied a 3 volt sine wave across the two wires at decreasing frequency loads (10 Hz for 30 min, 3 Hz for 15 min, 1 Hz for 7 min, and 0.5 Hz for 7 min). As noted by other investigators, giant vesicles formed best above its phase transition temperature.[26] Giant vesicles from 10 to 50 μm in diameters formed on the electrodes, and we cooled the vesicles to room temperature (~23 °C) at a cooling rate of ~0.28 °C/min. We harvested the vesicles in eppendorf vials and used them within a few days.

Micropipette Aspiration

Vesicles in an aqueous media were viewed with an inverted Nikon Diaphot 300 microscope (Nikon INC, Melville, NY) equipped with a 40X Hoffman modulation contrast objective (Modulation Optics, Greenvale, NY). The aqueous media containing the immersed 100 mM sucrose-filled vesicles varied from either 85 or 100 mM glucose with or without alcohol, depending on the type of experiments. Morphologies of DPPC vesicles were viewed in isotonic 100 mM glucose solutions. Flow-pipette MPA experiments on DPPC vesicles were performed in exterior 85 mM glucose solutions while flow-pipette MPA, area compressibility, and lysis tension experiments on DPPC/Chol vesicles were conducted in exterior 100 mM glucose solutions. The micropipettes used for holding vesicles were filled with a 100 mM glucose, 0.02 wt% albumin solution, and the tips of the micropipettes were coated with SurfaSil®. Micromanipulation of the vesicles and other details of the micropipette aspiration setup were thoroughly discussed in our previous paper,[14] and we note only changes in procedure here.

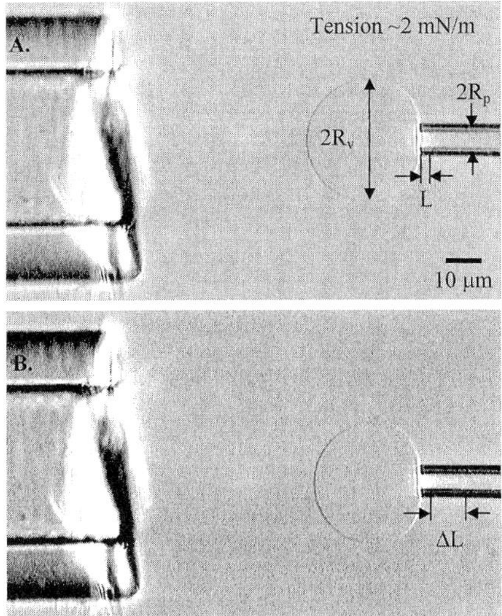

Figure 1 Microscope images of a weakly aspirated giant DPPC vesicle inside a micropipette. A flow pipette delivering a stream of an isotonic glucose media with 10 vol% ethanol is placed in front of the vesicle. **(A)** The initial length of the projection, L, before the flow is activated. **(B)** The projection length grew, ΔL, as the membrane underwent the $L_{\beta'}$-$L_{\beta}I$ phase transition. ΔL is proportional to the membrane area expansion, $(\Delta A/A_{0\%})_{exp}$.

We applied the micropipette aspiration technique developed by Evans and co-workers[18] to measure three mechanical properties of DPPC/Chol vesicles exposed to an alcohol/water mixture: membrane area expansion, area compressibility modulus, and lysis tension. We measured the area expansion by first weakly aspirating individual vesicles into a micropipette (7-9 μm in diameter). The aspirated vesicles (enclosing 100 mM sucrose) were suspended in an outside 85 mM glucose mixture where the difference in osmolarity swelled the vesicles into almost spherical shapes. Selected vesicles appeared clear (no trapped internal smaller vesicles), unilamellar, and ranged in size from 20 to 40 μm in diameter. Each aspirated vesicle was then prestretched beyond 7 mN/m for 1-2 seconds and relaxed back to a holding tension of ~2 mN/m. We next placed a flow pipette (100-150 μm in diameter) directly in front of the aspirated vesicle (see figure 1) to deliver a constant flow of an isotonic 85 mM glucose, alcohol/water mixture around the vesicle. The flow was driven by a syringe pump (KD Scientific Model 200, New Hope, PA) for constant velocities between 100 to 200 μm/s. The alcohol's interaction with the vesicle caused the vesicle's projection, L, inside the micropipette to elongate dramatically (see figure 1B) and for the spherical portion of the vesicle outside the micropipette to shrink in diameter. We video recorded this dynamic growth, ΔL, of the projection length and the decrease in the outer vesicle's radius until the projection length reached a plateau or the vesicle collapsed inside the micropipette. Collapse occurred quickly within 1-2 seconds. We subsequently analyzed the tapes to convert these vesicle's geometric parameters into an apparent area expansion, $(\Delta A/A_{0\%})_{\mathrm{exp}}=(A-A_{0\%})/A_{0\%}$ with the following equation with the inherent assumption that the volume of the aspirated vesicle remained constant.[27]

$$\left(\frac{\Delta A}{A_{0\%}}\right)_{\mathrm{exp}}=\frac{2\pi R_p \Delta L}{A_{0\%}}\left(1-\frac{R_p}{R_v}\right) \qquad (1)$$

We approximated the oblate shape of the aspirated vesicle as a sphere with an averaged radius, R_v. $A_{0\%}$ is the measured surface area of the vesicle before alcohol exposure, A is the surface area after exposure, and R_p is the radius of the micropipette. We confirmed that for vesicles exposed to a high ethanol concentration of 10 vol%, the volume of the individual aspirated vesicle (from measuring the projection growth, ΔL, and decreases in R_v) as a function of time until the projection reached a stable plateau was conserved. The differences in these measured time-elapsed volumes was less than or fell within the individual measured volume's uncertainty of 7%-14% for vesicles ranging from 20 to 40 μm in diameters with an optical resolution limitation of 0.5 to 1 μm.

The area compressibility modulus, K_{app}, represents the resistance to fractional membrane area dilation, $\alpha=(A-A_o)/A_o$, from an applied isotropic tension, τ, where A_o is the membrane area of the vesicle measured at an initial low tension state. We determined K_{app} for each giant DPPC/Chol vesicle by first weakly aspirating it inside a micropipette. Each vesicle was prestressed beyond 7 mN/m for 1-2 seconds to smooth out any folds or tethers in the membrane and relaxed back to a desired low tension of ~1 mN/m. The applied suction pressure, ΔP, holding the vesicle was next increased in a stepwise manner to induce an incremental deformation change, ΔL, in the vesicle's projection, L, inside the micropipette. For each pressure change, ΔP was held constant for at least five seconds to ensure that L has reached equilibrium. We recorded these stress-strain measurements onto videotapes and later analyzed the vesicle images to measure α and τ. α was calculated with equation 1 whereas τ was computed from approximating the oblate shape of the vesicle as a sphere with an averaged radius, R_v:[27]

$$\tau = \frac{\Delta P \, R_p}{2\left(1 - \frac{R_p}{R_v}\right)} \quad (2)$$

The plot of τ versus α exhibited a linear relationship, and the slope of this line by convention is known as the apparent area compressibility modulus, K_{app}, i.e. $\tau = K_{app}\alpha$.[28] Lysis tension, τ_{lys}, is the highest mechanical tension the vesicle can sustain before rupturing or collapsing. We measured τ_{lys} of DPPC vesicles in 100 mM glucose at a pulling rate of 0.1 mN/m/s and in 85 mM glucose at a pulling rate of 0.5 mN/m/s. DPPC/Chol vesicles were measured in 100 mM glucose mixtures with and without alcohol at pulling rates of 0.2 mN/m/s and 0.7 mN/m/s respectively.

Results and Discussion

Surface Morphology and Membrane Failure of DPPC Vesicles

A major experimental objective of this work was to subject giant DPPC vesicles to flows of ethanol solutions while being held by micropipette aspiration. The subsequent increases in membrane projection lengths would allow us to quantify membrane area expansion $(\Delta A/A_{0\%})_{exp}$ accompanying the $L_{\beta'}$-$L_{\beta}I$ phase transition. However, in order to successfully carry out this experiment, aspirated DPPC vesicles must have a smooth surface so that $(\Delta A/A_{0\%})_{exp}$ represents the $L_{\beta'}$-$L_{\beta}I$ phase transition rather than smoothing of sharp membrane contours, aspirated vesicles must be robust enough to stay intact under an applied tension, and DPPC vesicles must still remain as stable vesicles after the $L_{\beta'}$-$L_{\beta}I$ phase transition. Therefore, we characterized giant DPPC vesicles in terms of surface morphology, general stability and morphology after addition of ethanol, and lysis tension. This allowed us to develop a feasible experimental approach for the ethanol flow experiments described in the next section.

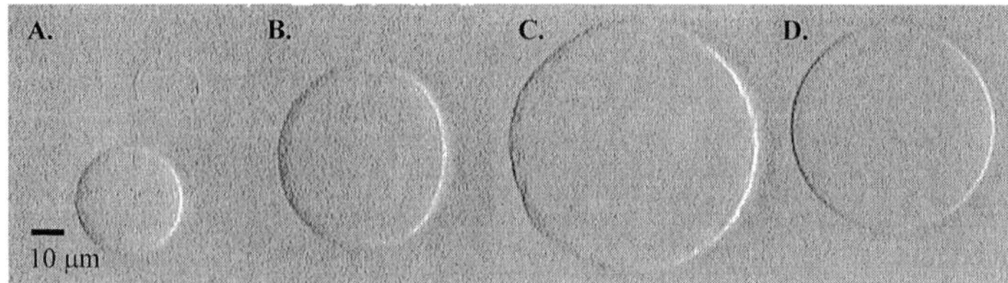

Figure 2 Micrographs of gel phase giant vesicles observed with a 40X Hoffman contrast objective in an isotonic glucose media without alcohol. **(A)** Round DPPC vesicle (bottom) contrasted with an almost transparent ghost DPPC vesicle (top). **(B)** Oblate DPPC vesicle. **(C)** Multifaceted DPPC vesicle. **(D)** Spherically round DPPC/25 mol% Chol vesicle.

DPPC vesicles enclosing 100 mM sucrose in the $L_{\beta'}$ phase were added to an isotonic 100 mM glucose solution without alcohol. We observed a large population of vesicles with a distributed morphology (see figure 2) ranging from completely spherical to multifaceted shapes that most probably depended on the original surface area/volume ratio of the vesicles created during electroformation. We also observed that the percentage of transparent "ghost" vesicles was larger than typically observed in the case of fluid-phase vesicles. When we added the same electroformed DPPC vesicles to an isotonic 100 mM glucose solution containing 10 vol% ethanol (above the $L_{\beta'}$-$L_{\beta}I$ threshold concentration of ~7 vol%), we noticed vesicles displaying the same range of morphology, but a greater number of multifaceted vesicles were seen, presumably due to the induction of the interdigitated state. With or without ethanol, the surfaces of the vesicles appeared rigid and did not thermally fluctuate at room temperature (as expected for the rigid $L_{\beta'}$ and $L_{\beta}I$ phases). These morphological observations agree well with descriptions of gel phase giant vesicles composed of dimyristoyl phosphatidylcholine (DMPC, diC14:0 PC), a shorter chain analog of DPPC.[27,29]

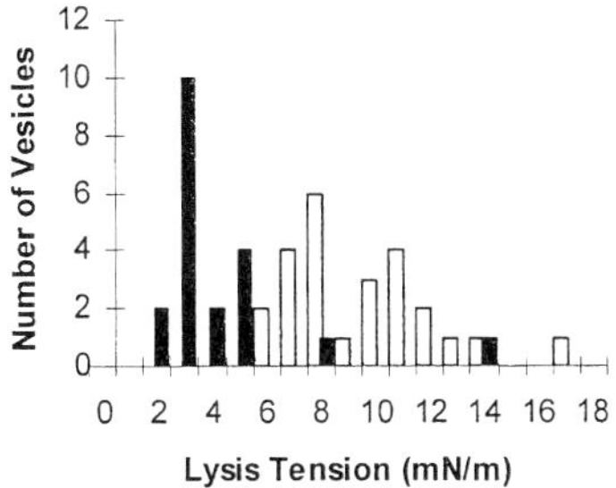

Figure 3 A histogram of the lysis tension values of pure DPPC vesicles mechanically pressurized at a slow pulling rate of 0.1 mN/m/s (black, n=20 vesicles) and high pulling rate of 0.5 mN/m/s (white, n=25 vesicles).

When we aspirated DPPC vesicles at an initial tension of ~1-2 mN/m in an isotonic 100 mM glucose solution without ethanol until membrane failure, the projection length grew monitonically with applied tension, we presume primarily due to the smoothing out of shape deformation. Afterward, the vesicles typically crumpled like "paper sacks" before completely being drawn inside the micropipette within 1-2 seconds. Gel phase lipids are predicted to have higher lysis tension than fluid phase lipids because of their close packing and strong Van der Waals interactions.[27] However, as shown in figure 3 and 10, we found that these vesicles have lysis tension ranging from 1 to 16 mN/m with the majority of aspirated vesicle rupturing around 4 mN/m at a pulling rate of 0.1 mN/m/s. This value can be compared to a recent measured lysis tension value of 9-10 mN/m for fluid phase PC vesicles measured at the same pulling rate of 0.1 mN/m/s.[30] The lower lysis tension value for gel phase DPPC vesicles implies that the vesicles were riddled with surface defects that served as sources for pore nucleation. Stochastic formation of pores, which represents a balance between membrane tension τ and line tension λ, is believed to cause membrane failure if they reach an energetically unstable size.[31] When DPPC vesicles were cooled through the main fluid-gel phase-transition temperature after electroformation, permanent surface defects such as disclinations, holes, or grain-boundaries could form. It is likely that the edges in multifaceted vesicles were grain-boundaries and that ghost vesicles resulted from exchange of solute through surface defects.

Attempts to reduce surface defects by annealing were unsuccessful. Specifically, following the protocols of Kim,[32] we annealed freshly formed gel-phase DPPC vesicles by cycling three times between room temperature and 34 °C (around the ripple-phase transition temperature of 35 °C, but away from the main transition temperature of 41 °C[25]). However, when subjected to micropipette aspiration at a pulling rate of 0.1 mN/m/s, these vesicles continued to rupture around 4-5 mN/m. Perhaps the annealing conditions were insufficient to reduce significantly the surface defect density or that a permanent number of defects always existed in the constrained round or multifaceted vesicles. Therefore, we did not bother to anneal DPPC vesicles used in subsequent MPA experiments. Although the pulling rate of 0.1 mN/m resulted in DPPC vesicles lysing around 4 mN/m, we observed DPPC vesicles lysed at higher tensions (~9 mN/m) as expected when the pulling rate was increased to 0.5 mN/m/s (figure 3 and 10) because measured lysis tension increases with increasing pulling rate.[33,34]

Ethanol Induced Membrane Area Expansion and Threshold Concentration

We monitored projection length growth of DPPC vesicles held by micropipette suction as the vesicles were exposed to flows of isotonic alcohol/water mixtures. Steps were taken in order to assure that the projection growth was entirely due to a phase transition and not to shape-induced deformations under mechanical tension.[27] First, DPPC vesicles enclosing 100 mM sucrose were added to an 85 mM glucose media to inflate them into quasi-spheres without rupturing them. Next, upon aspiration, these vesicles were prestretched beyond 7 mN/m over a very short period of time, 1-2 seconds. Each vesicle's projection length grew inside the micropipette and the outside contour of the vesicle became smoother. The vesicles often still appeared oblate, but further pressurization to increase sphericity (and remove all shape deformations) only resulted in rupture (due to their inherently low lysis tension). The pressure was then relaxed to ≤ or ~2 mN/m wherein the projection length did not decrease and the general shape of the vesicle did not change (as monitored for at least 15 minutes.)

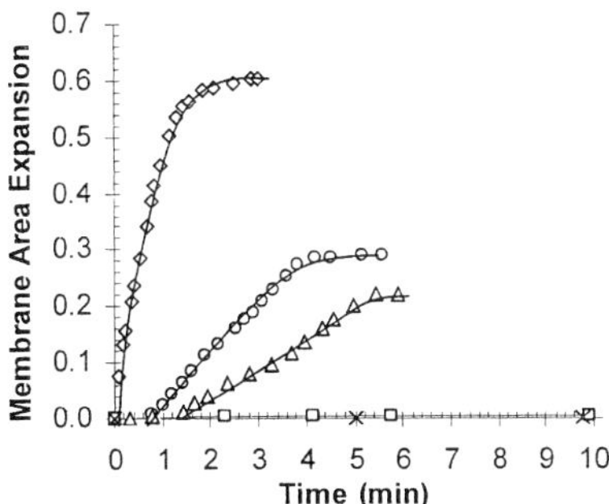

Figure 4 Instantaneous membrane area expansion $(\Delta A/A_{0\%})_{exp}$ of individual vesicles exposed to a flow pipette delivering a constant isotonic ethanol/water mixture stream at specific ethanol concentration (vol): 0% (—×—), 5% (—□—), 8% (—△—), 10% (—○—), and 15% (—◇—). The plateau values of the membrane expansion are reported in figure 5 as equilibrium expansion of the $L_{\beta'}$-$L_{\beta}I$ phase transition. Line fitting is intended as a guide for the eye.

We exposed aspirated DPPC vesicles to a series of ethanol/water mixtures containing 85 mM glucose: 0, 5, 6, 7, 8, 10, 15, and 20 vol% ethanol. The equivalent molar concentrations are 0, 0.85, 1.02, 1.19, 1.36, 1.70, 2.56, and 3.41 M, respectively. For vesicles held at a low tension of ≤1 mN/m and exposed to these isotonic ethanol/water mixtures above 7 vol%, we observed that the oblate shape of the vesicles became wrinkly and folds or creases formed along the surface (presumably from the $L_{\beta'}$-$L_{\beta}I$ phase transition) while the projection length remained unchanged. The low tension of ≤1 mN/m was not sufficient on the timescale of the experiment to pull the excess surface area into the micropipette presumably because of the viscous nature of the interdigitated phase. When the holding tension was larger, ~2 mN/m, we observed that the projection length quickly grew upon exposure to ethanol concentration above 7 vol% as the excess surface area was smoothed out, and the portion of the vesicle external of the pipette maintained its smooth surface. However, at this holding tension, the vesicles were more prone to collapse before the projections reached stable plateaus (possibly due to the fact that thinner membranes have inherently lower lysis tensions[35]). Only runs where stable projections were reached (for example see figure 4) were used for reporting equilibrium $(\Delta A/A_{0\%})_{exp}$. Calculated equilibrium $(\Delta A/A_{0\%})_{exp}$ values from the growth in the projection length and the decrease in the radius of the outer portion of the aspirated vesicle were due to areal expansion and not due to internal volume change of the vesicle (as discussed in the Material and Methods section). For the equilibrium runs, the projections reached stable plateaus within 3-6 minutes and generally remained unchanged until spontaneous rupture occurred or the flow was manually removed. In the presence of an isotonic glucose solution flow (no ethanol), we observed practically no change in $(\Delta A/A_{0\%})_{exp}$ as shown in figure 4.

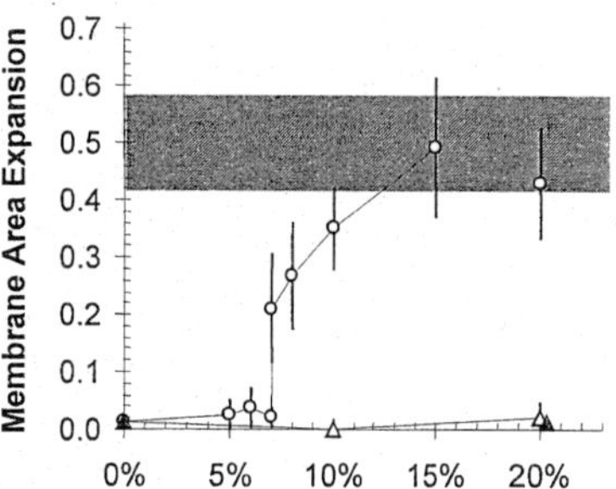

Figure 5 Equilibrium membrane area expansion $(\Delta A/A_{0\%})_{exp}$ vs. ethanol concentration for vesicles with 0 mol% Chol (—o—), 10 mol% Chol (—△—), and 25 mol% Chol (▲). The shaded region represents the expected membrane area when the complete surface fully interdigitates. For 0 mol% Chol at 0%, 5%, 6%, 7% bottom, 7% top, 8%, 10%, 15%, and 20% ethanol (vol), n = 8, 6, 13, 5, 5, 7, 9, 6, and 9, respectively where n is the number of vesicles averaged. For 10 mol% Chol at 10% and 20% ethanol (vol), n = 4 and 6 while for 25% Chol at 20% ethanol (vol), n = 4. Bar indicates one standard deviation.

Equilibrium $(\Delta A/A_{0\%})_{exp}$ values of vesicles with stable projection lengths are reported in figure 5. We observed a sharp threshold occurring at 7 vol% ethanol. At 6 vol%, no area expansion was observed; above 7 vol%, $(\Delta A/A_{0\%})_{exp}$ increased monotonically from 20% to 50% with increasing ethanol concentration. The area expansion at 7 vol% ethanol was bimodal with some vesicles displaying no change in membrane area while others expanded around 20%, and the bimodal distribution appeared invariant with holding tension ranging from 1 to 2 mN/m. This threshold ethanol concentration of 7 vol% for induction of area expansion of giant unilamellar DPPC vesicles concurs with previously reported values of 6.3 vol% for induction of the $L_{\beta'}$-$L_{\beta}I$ transition as measured by DSC of multilamellar vesicles.[36] Measurement by X-ray diffraction of multilamellar DPPC vesicles revealed that upon full interdigitation, the membrane thickness of 4-5 nm decreased by 1.6 nm.[1] Similarly, AFM studies on single supported DPPC bilayer showed that the membrane thickness of 5.7 nm decreased by 1.9 nm[37] when interdigitation occurred. For this membrane decrease of ~33 ± 4%, we would expect that when the membrane lipid fully interdigitates, the membrane area should expand around 50 ± 8% (with the assumption that the volume of the hydrocarbon core of the lipid membrane is generally conserved as measured by wide angle reflection of DPPC multibilayers in a large range of ethanol/water mixtures[6]). Though we clearly see a sharp difference in area expansion beyond 7 vol% ethanol, a strong indication that interdigitation occurred, the values between 7 and 15 vol% are below the predicted value of 50% for full interdigitation. There are two possible rationales. The complete membrane of the giant aspirated vesicle became partially interdigitated. This seems thermodynamically implausible. The incomplete packing of the rigid lipid chains would signify a considerable loss of van der Waals interaction energy, and the voids created between the lipid hydrocarbon tails for exposure to water would be highly energetically unfavorable. The more probable rationale is that instead of the entire membrane being fully interdigitated, only large regions or domains of DPPC lipids in the membrane were fully interdigitated while the rest of the membrane remained in the noninterdigitated state. This view is consistent with AFM studies that showed noninterdigitated and interdigitated DPPC domains coexisted on the mica surface in the presence of ethanol below 20 vol% ethanol.[37] At 20 vol% ethanol, the area fraction of the interdigitated phase was 100%. Vierl et al.[6] drew a phase diagram of DPPC in various water-ethanol mixtures (based on the combination of calorimetric, fluorescence, dynamic light scattering, and X-ray diffraction data) that showed both the noninterdigitated and interdigitated phases can coexist for ethanol concentration higher than ~7 vol% in a temperature range of 18 and 25 °C. Other supporting evidence for the coexistence of the fully interdigitated and non-interdigitated phases are found in fluorescent studies of multilamellar DPPC vesicles in the presence of ethanol,[12] X-ray diffraction of dihexadecylphosphatidylcholine (DHPC)/cholesterol vesicles,[13] and high-pressure light transmission of DPPC multilamellar vesicles in the presence of the anesthetic tetracaine.[38] Consequently, figure 5 shows that the size of the fully interdigitated region grew proportionally with increasing ethanol concentration from 7 to 15 vol%. At 7 vol%, about half of the membrane was interdigitated, and at 15 vol%, the whole membrane became interdigitated.

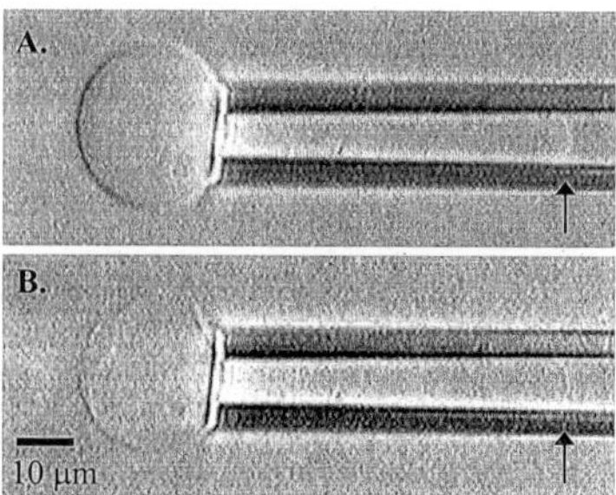

Figure 6 Images of an aspirated DPPC vesicle after being exposed to a flowing stream of an isotonic glucose solution containing 10 vol% ethanol for 7 minutes. **(A)** The projection has reached a stable plateau after undergoing the $L_{\beta'}$-$L_{\beta}I$ phase transition with a $(\Delta A/A_{0\%})_{exp}$ value of 30% with no detectable internal volume loss in the vesicle within experimental error. **(B)** After the flow pipette was removed, the projection length remained unchanged for over 35 minutes, indicating a kinetically trapped interdigitated state. Reducing the holding tension from 2 to 0.2 mN/m caused the surface of the vesicle outside the pipette to wrinkle. Arrows indicate the edge of the projection length.

Kinetically Trapped Ethanol Membrane Area Expansion

When the projection growth reached a plateau during exposure to an isotonic ethanol/water mixture, the plateau generally remained stable for a very short time (< 1 minute). On few occasions, some plateaus were stable for a longer period (3 to 5 minutes). For these vesicles, we moved the flow pipette away from the vesicle and stopped the flow. We expected the projection length to slowly retract as ethanol molecules in the vesicle's interior, in the membrane, and outside media diffused away and caused the interdigitated membrane to undergo a phase transition back to the gel $L_{\beta'}$ phase. This expectation was based upon our previous work with fluid phase PCs in which we demonstrated, by the same MPA-flow technique, that the area expansion caused by exposure to short-chain alcohols was reversible.[14] Instead, we observed that for interdigitated DPPC vesicles with 10 vol% ethanol/water media removed, the projection length remained unchanged (see figure 6). Reducing the holding tension from ~2 mN/m to ~0.2 mN/m should induce the projection length to withdraw; instead only the surface of the DPPC vesicle outside the micropipette relaxed to its corrugated shape (figure 6B). Apparently, the DPPC vesicle was kinetically trapped in the interdigitated state though certainly the ethanol concentration in the surrounding aqueous media was nearly zero as sufficient time (minutes) had passed for the small amount of added ethanol to disperse into the chamber. This ostensibly irreversible effect had been previously observed using DSC that showed the $L_{\beta}I$-$L_{\beta'}$ phase transition for PCs can be very slow, requiring weeks to reach completion.[5] AFM studies showed the supported interdigitated domains remained stable for a period of days after the ethanol was removed from the bathing solution.[37] However, heating the interdigitated domains above the main phase transition restored the noninterdigitated bilayer. Interestingly, this effect of an irreversible phase transition is being applied to construct a multicompartmental aggregate of tethered vesicles encapsulated within a large bilayer vesicle, known as a vesosome.[11]

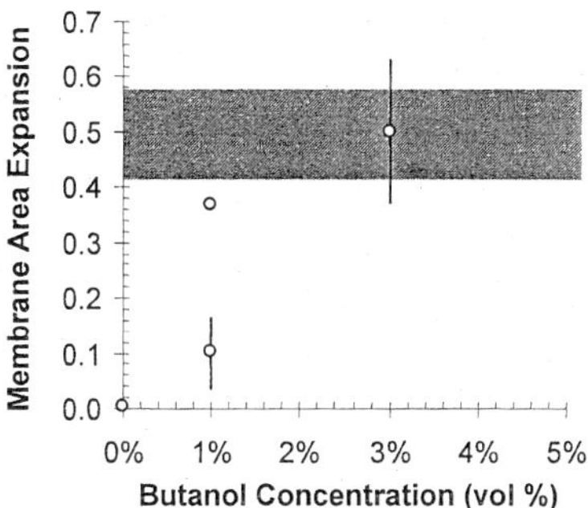

Figure 7 Equilibrium membrane area expansion $(\Delta A/A_{0\%})_{exp}$ vs. butanol concentration. The shaded region represents the expected membrane area when the complete surface fully interdigitates. For 1% (bottom), n=5; 1% (top) n=2; and 3% n=6 where n is the number of vesicle averaged. Bar indicates one standard deviation except at 1% where both values are equal to 0.37.

Butanol Membrane Expansion, Threshold Concentration, and Traube's Rule

Alcohols of longer chain length than ethanol have been shown to induce the interdigitated state in DPPC membranes at much lower concentrations.[39] Indeed, our MPA-flow experiments confirmed that giant DPPC vesicles exposed to an isotonic 1 vol% butanol/water solution was sufficient to induce the interdigitated state as shown in figure 7. A bimodal distribution was observed at 1 vol% butanol where some vesicles showed $(\Delta A/A_{0\%})_{exp}$ value of ~10% while others exhibited ~35%. A higher butanol concentration of 3 vol% caused all vesicles to expand around 50%, signifying that the surfaces were completely interdigitated. The observation of interdigitation at 1 vol% butanol for giant unilamellar vesicles agrees favorably with previously reported values of 1.2 vol% for the threshold concentration necessary for interdigitation measured by titration calorimetry and DSC of multilamellar vesicles.[40]

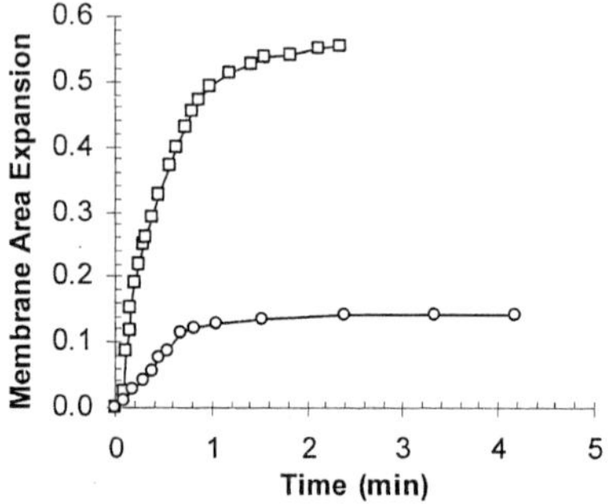

Figure 8 Instantaneous membrane area expansion $(\Delta A/A_{0\%})_{exp}$ of individual vesicles exposed to a flow delivering a butanol/water mixture at 1 vol% (—o—) and 3 vol% (—□—).

A comparison of figure 8 to figure 4 shows the timescale for $L_{\beta'}$-$L_{\beta}I$ phase transition was considerably shorter for vesicles exposed to butanol than ethanol near their respective threshold concentrations. Projection lengths of vesicles exposed to ethanol equilibrate within 3-4 minutes while vesicles exposed to butanol quickly equilibrated within 1 minute. Presumably, butanol's enhanced hydrophobicity from its longer hydrocarbon tail led to enhanced partitioning of the alcohol molecules into the DPPC membrane.[14,40]

Previously reported threshold concentrations for the short-chain alcohol family (methanol, ethanol, propanol, and butanol) in molarity units were 2.5, 1.02, 0.33, and 0.13 M (or in volumetric units are 10.1, 6.3, 2.5, and 1.2 vol%).[39,40] Note that the difference between each number is approximately a factor of three, signifying that for each additional alcohol CH_2 group, the concentration required to induce the $L_{\beta'}$-$L_{\beta}I$ phase transition was reduced roughly by a factor of three. This finding follows Traube's rule of interfacial tension reduction of a hydrophobic-water/alcohol interface due to alcohol adsorption at the interface, i.e. for each additional CH_2 group, the concentration required to reach the same interfacial tension was reduced by a factor of three.[41] As we discussed in our previous work on the adsorption of short-chain alcohols into fluid phase 1-stearoyl, 2-oleoyl phosphatidylcholine (SOPC) membrane, enhanced partitioning of alcohol molecules into the membrane with increasing chain length is at the root of the Traube's rule effect.[14]

Hence, it is likely that the $L_{\beta'}$-$L_{\beta}I$ phase transition is triggered when the surface alcohol concentration in the $L_{\beta'}$ state reaches a single critical value as noted by other investigators,[42,43] and the bulk alcohol concentration required to reach that value decreases by a factor of three for each additional CH_2 group. Mole fractions or partition coefficients of alcohol adsorption in the $L_{\beta'}$ state are difficult to measure and unavailable,[40] but Rowe (1985) showed the critical amount in the fluid phase DPPC was ~0.10 mole fraction to induce the L_{α}-$L_{\beta}I$ phase transition for methanol, ethanol, and propanol.

Impact of Cholesterol on Ethanol Induced Membrane Area Expansion

The incorporation of cholesterol into the gel phase DPPC membrane is known to inhibit the induction of the $L_{\beta'}$-$L_{\beta}I$ phase transition above the alcohol threshold concentration.[12,44] To further investigate this phenomenon, we formed giant vesicles containing DPPC and 10 mol% or 25 mol% Chol for use in MPA-flow studies similar to those performed using pure DPPC vesicles. These vesicles appeared spherically round (figure 2D). No multifaceted vesicles were observed, and at various times, the membrane surface thermally undulated showing that surface rigidity was removed. Upon aspiration into a micropipette, the contour of the vesicle outside the pipette always appeared completely round. The vesicle's projection inside the pipette increased linearly with increasing suction pressure, and the process showed a perfectly reversible (elastic) recovery upon decreasing membrane tension (see figure 9). According to the DPPC:Chol phase diagram, DPPC and 10 mol% Chol should exist in between the gel phase and gel – liquid-ordered phase coexistence regime at room temperature. The DPPC and 25 mol% Chol bilayer exists only in the liquid-ordered phase which exhibits simultaneously fluid and solid like properties.[45] Our observations are consistent with the presence of a liquid-ordered phase in which free moving lipids and cholesterol eliminates any long-lived surface defects and surface rigidity.[46]

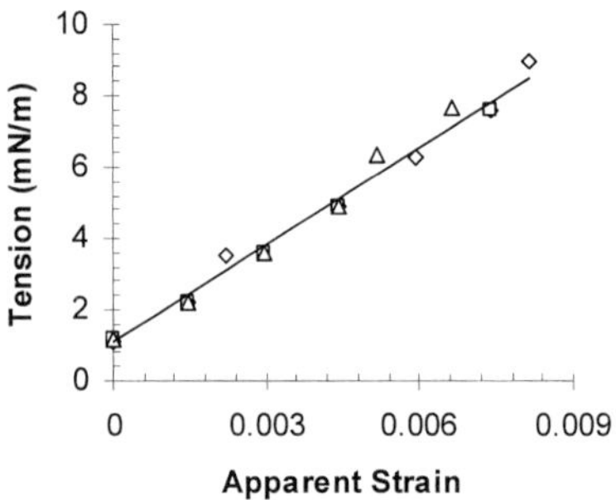

Figure 9 Tension-strain plot of DPPC/10 mol% Chol showing its reversibility when tension was increased (◇) to a high value, decreased (□) back to its initial holding tension, and increased until the vesicle ruptured (△) at a pulling rate of 0.2 mN/m/s. The K_{app} value is 930 mN/m.

DPPC/Chol vesicles aspirated with a holding tension of ~2 mN/m in a 100 mM glucose solution were exposed to an isotonic ethanol/water flow stream with specific ethanol volume concentration of either 0%, 10% or 20%. We observed no increase in the projection length, indicating that induction of the $L_{\beta'}$-$L_{\beta}I$ phase transition was inhibited in vesicles containing 10 mol% or 25 mol% cholesterol (see figure 5). Komatsu and Rowe (1991)[12] showed with flurophores covalently coupled to the ends of the DPPC acyl chains that the induction was completely prevented by the presence of 20 mol% Chol in 12 vol% ethanol. Therefore, our experiment shows that an even lower concentration (10 mol%) will abolish the transition in DPPC/Chol bilayers. Our value is consistent with Lu et al.[47] who showed that at 25 °C, 10 mol% Chol incorporated in 30 mol% 16:0 lysophosphatidylcholine (LPC) / DPPC vesicles inhibits the $L_{\beta'}$-$L_{\beta}I$ phase transition that would otherwise had been induced by the 30 mol% presence of LPC. DSC measurements revealed that the noninterdigitated bilayer structure of 30 mol% LPC / DPPC with 10 mol% Chol incorporation was more thermodynamically stable than the interdigitated phase of 30 mol% LPC / DPPC as evidenced by the bilayer having a significant lower main phase transition temperature ($L_{\beta}I$ - L_{α}) and phase transition enthalpy. A lower amount of 5 mol% Chol has been reported to eliminate completely the interdigitated phase for the DHPC gel phase systems.[13]

The Influence of Ethanol on Mechanical Properties of DPPC/Cholesterol Vesicles

Unlike DPPC vesicles, vesicles formed of DPPC/Cholesterol did not lyse at low tensions and had smooth membrane contours. Therefore, using micropipette aspiration, we could obtain sufficient tension, τ, vs. areal strain, α, data to determine the influence of ethanol on area compressibility, K_{app}, and lysis tension, τ_{lys}, of DPPC/Chol bilayers. DPPC/Chol vesicles containing 10 mol% or 25 mol% Chol formed in 100 mM sucrose were placed in an isotonic 100 mM glucose solution containing 0%, 10%, or 20% ethanol (vol). Subsequent micropipette aspiration resulted in reversible linear growth in the projection length as the suction pressure was increased (figure 9). Following common micropipette convention,[28] the slope of this line is the apparent area compressibility moduli K_{app}, and average values are listed in table 1.

Table 1. Area compressibility moduli K_{app} (mN/m) obtained for giant unilamellar vesicles by the micropipette aspiration method

Cholesterol content in DPPC/Cholesterol Vesicle (mol%)	Ethanol Concentration (vol%)		
	0	10	20
10	900 ± 240 (10)	810 ± 160 (9)	760 ± 290 (4)
25	1200 ± 220 (9)	-	750 ± 90 (11)

*Parentheses represented number of vesicles measured. The difference between the K_{app} values of vesicles with 10 mol% and 25 mol% Chol in 0 vol% ethanol is P=0.013 (as evaluated by Student's t-test) whereas the difference between the K_{app} values of vesicles with 25 mol% Chol in 0 and 20 vol% ethanol is P= 0.0002.

The K_{app} values of DPPC vesicles with 10 and 25 mol% cholesterol content of 900 ± 240 and 1200 ± 220 mM/m, respectively, are higher but within order of the value of 740 ± 100 mN/m reported for DPPC vesicles with 40 mol% cholesterol.[48] Direct measurement of K_{app} values for DPPC bilayer membrane with or without cholesterol appeared to be limited in the literature although the K_{app} value for the shorter-chain analog DMPC exists as well as K_{app} values for fluid-phase SOPC bilayer with cholesterol.[26,27,49] Evans and Needham[27] listed the K_{app} value for the pure gel-phase DMPC as 860 ± 140 mN/m. This predicted value's similarity to our measured K_{app} values with the addition of either 10 or 25 mol% cholesterol is consistent with current knowledge that cholesterol, when added to the membrane, has the unusual property of maintaining strong membrane cohesion while simultaneously softening the membrane by abolishing surface rigidity.[49] For example, for fluid-phase SOPC vesicles, inclusion of 50 mol% cholesterol had been shown to quadruple the K_{app} value from 200 to 800 mN/m.[49,50]

When we pressurized DPPC/Chol vesicles to measure their K_{app} values in a 10% or 20% (vol) ethanol/water mixture, the DPPC/Chol vesicles behaved elastically, similar to their counterparts in an ethanol-free/water mixture. However, DPPC/10 mol% Chol in 20 vol% ethanol behaved elastically only below 8.3 ± 1.4 mN/m and yielded plastically above this value (discussed in next section). As shown in table 1, the average K_{app} values for DPPC/10 mol% Chol of 810 ± 160 mN/m in 10 vol% ethanol and 760 ± 290 mN/m in 20 vol% ethanol (in the elastic regime) are comparable to 900 ± 240 mN/m in 0 vol% ethanol. However, the average K_{app} value for DPPC/25 mol% Chol of 750 ± 90 mN/m in 20 vol% ethanol is significantly lower than 1200 ± 220 in 0 vol% ethanol. Apparently, the ethanol's interactions with DPPC/25 mol% Chol membrane partially decreased the membrane cohesion, and this implies that perhaps adsorption of ethanol molecule at the bilayer-water interface has reduce the surface energy per monolayer of the membrane. The difference in the K_{app} values further suggests that during the MPA-flow experiments, we might observe a small $(\Delta A/A_{0\%})_{exp}$ value from mechanical deformation as the K_{app} decreases from ethanol exposure. The following calculation will show that $(\Delta A/A_{0\%})_{exp}$ would be too minuscule to notice. The expected area strain $\alpha(K_{app})$ can be estimated with this equation:

$$\alpha(K_{app}) = \tau\left(\frac{1}{K_{app,\ alcohol}} - \frac{1}{K_{app,\ water}}\right) \tag{3}$$

where τ is the holding tension set at 2 mN/m, $K_{app,\ alcohol}$ is the average area compressibility modulus of vesicles in a specific alcohol/water mixture, $K_{app,\ water}$ is the average area compressibility modulus of vesicles in an aqueous mixture without alcohol. The $\alpha(K_{app})$ computed is less than 0.0009, and indeed no area expansion was seen, as evidenced in figure 5.

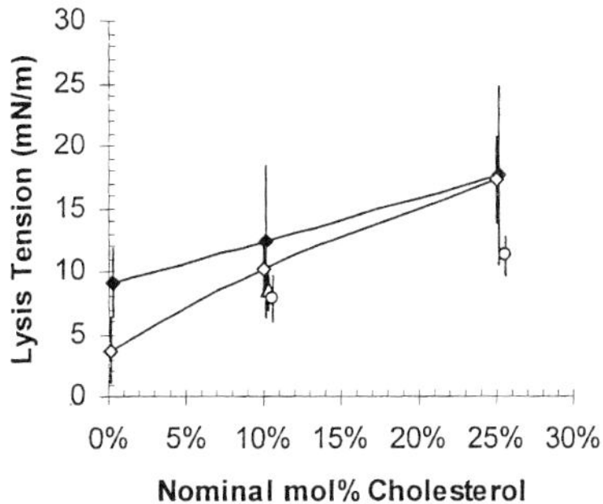

Figure 10 Lysis tension τ_{lys} of DPPC/Chol vesicles suspended in various ethanol concentration (vol %) and subjected to different pulling rates: 0 vol % ethanol, 0.5-0.7 mN/m/s (—◆—); 0 vol %, 0.1-0.2 mN/m/s (—◇—); 10 vol%, 0.2 mN/m/s (△); and 20 vol %, 0.2 mN/m/s (○). The average values are based on 10-20 vesicles. Bar indicates one standard deviation.

In a separate set of experiments, we performed micropipette aspiration in order to determine the lysis tension values of DPPC vesicles containing 10% or 25% (mol) cholesterol in 0%, 10%, or 20% (vol) ethanol (figure 10). For vesicles in 0% ethanol subjected to a pulling rate of 0.5-0.7 mN/m/s, inclusion of 10 and 25 mol% Chol increased the average lysis tension value from 9 ± 3 mN/m at 0% Chol to 13 ± 6 and 18 ± 7 mN/m, respectively. At the medium pulling rate between 0.1 to 0.2 mN/m/s, 10 and 25 mol% Chol substantially raised the lysis tension value from 4 ± 3 mN/m at 0% Chol to 10 ± 2 and 17 ± 3 mN/m, respectively. Surprisingly, the lysis tension values at both 10 and 25 mol% cholesterol showed no strong dependence on pulling rates from 0.1 to 0.7 mN/m/s. Extrapolation of the lines in figure 10 to 40% cholesterol results in an average lysis tension value of approximately 24 mN/m which compared favorably with reported value of 23 ± 3 for DPPC/40 mol% Chol.[48] Cholesterol toughened the surface of the DPPC/Chol vesicles by plausibly ameliorating surface defects and maintaining high membrane cohesion.

Figure 10 shows the lysis tensions of DPPC vesicles with 25 mol% Chol in 20 vol% ethanol were significantly reduced compared to systems without ethanol, while the lysis tensions of DPPC/10 mol% Chol were slightly reduced. This trend is in good agreement with the trend in K_{app} values of DPPC/Chol vesicles as shown in table 1. This is to be expected since the K_{app} value is linearly related to the lysis tension assuming a constant value for the areal strain at rupture. The areal strains at rupture for DPPC vesicles with 10 and 25 mol% Chol are 0.009 ± 0.004 and 0.012 ± 0.004, respectively, and remain unchanged in the presence of ethanol.

Cholesterol and Pressure-Induced Interdigitation

Above a yield tension of 8.3 ± 1.4 mN/m (n=4 vesicles, close to its lysis tension of 7.9 ± 1.8 mN/m), we observed that DPPC/10 mol% Chol vesicles in 20 vol% ethanol yielded plastically over minutes before the membrane collapsed within 1 to 2 seconds (see figure 11A). To ensure that the large growth in the projection was not simply due to lysis of the vesicle (where the vesicle would vanish away within 1-2 seconds) or volume change from pore formation, we held each pressure increment for ~1 minute (see figure 11B). No stable plateaus of the plastic deformation were ever seen, but the measured membrane area growth that occurred over minutes before collapsed varied from 7% to 20%. This showed that although 10 mol% Chol inhibited the induction of the $L_{\beta'}$-$L_{\beta}I$ phase transition in 20 vol% ethanol, the effect was negated when the mechanical tension (from an applied suction pressure) exceeded a critical value. We then verified this finding by performing MPA-flow experiments. As shown in figure 5, no membrane expansion occurred when DPPC/10 mol% Chol vesicles, held at a tension of 2 mN/m, were exposed to 20 vol% ethanol for a long period of time (over 8-10 minutes). However, when the holding tension was increased to ~9 mN/m, on a few occasions, we observed a large $(\Delta A/A_{0\%})_{exp}$ value, on average of 10%, before the vesicle disappeared (data not shown).

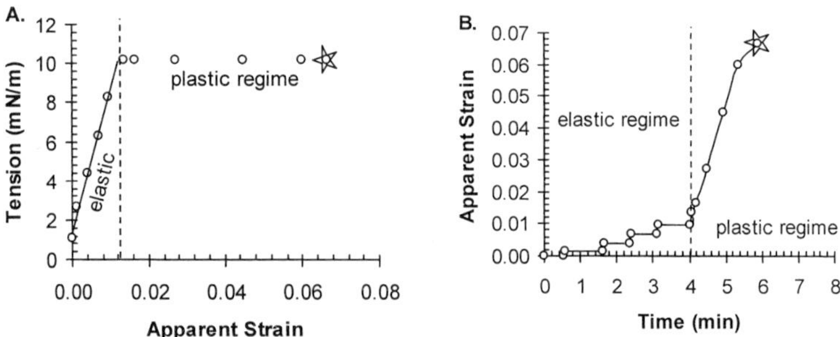

Figure 11 Tension-strain plot of a DPPC/10 mol% Chol vesicle in 20 vol% ethanol. At the yield tension of 10.2 mN/m, the $L_{\beta'}$-$L_{\beta}I$ phase transition was induced. **(A)** From the slope of the line up to the yield tension, K_{app} is 720 mN/m. **(B)** The strain rate is 0.003 min^{-1} at a pulling rate of 0.04 mN/m/s for the first 4 minutes where a holding time of ~1 minute followed each pressure increment. The star symbol represents the point at which the vesicle collapsed.

High pressure-induced interdigitation of saturated PCs in aqueous solutions are well documented in literature out of interest in understanding pressure-anesthetic antagonism, the adaptation of marine organisms to extreme depths, and the sterilization in food processing by high pressure.[3,51-53] Yet, this is the first reporting of low pressure-induced interdigitation of membranes. A high hydrostatic pressure P of ≥70 MPa was generally necessary to induce the $L_{\beta'}$-$L_{\beta}I$ phase transition in DPPC membranes.[54] The yield tension of 8.3 ± 1.4 mN/m for our DPPC/10 mol% Chol system in 20 vol% ethanol is equivalent to a low hydrostatic pressure of 4 MPa based on the formula $P=\tau/H$.[55] H is the effective size of the lipid molecule in the normal direction to the lamella surface: $H^{\beta}=2l\cos\theta$ in the $L_{\beta'}$ state, $H^{in} \approx l$ in the $L_{\beta}I$ state where l is the length of the lipid chain $l = n \times 1.27 \times 10^{-10}$ m; n is the number of carbon groups; θ is the chain tilt in the $L_{\beta'}$ state. A recent theoretical model based on membrane mechanical properties and consistent with experimental results has been proposed by Levadny et al.[55] for PCs in an aqueous solution without alcohol in a high hydrostatic pressure environment (>70 MPa). They discussed that the induced mechanical tension from the applied pressure in the lipid heads region is considerably relieved by a phase transition from either the L_{α} or $L_{\beta'}$ to the interdigitated $L_{\beta}I$ phase. Our observation of a much lower pressure of 4 MPa for our DPPC/10 mol% Chol system in 20 vol% ethanol is most likely due to the presence of a high concentration of ethanol. Ethanol and pressure can work cooperatively to induce interdigitation, as described by Zeng and Chong.[56]

Conclusions

In this work, we applied micropipette aspiration to giant DPPC unilamellar vesicles, thus enabling direct characterization of the areal expansion of the gel ($L_{\beta'}$) phase induced by exposure to short-chain alcohols: ethanol and butanol. Areal expansion began at a threshold ethanol concentration of 7 vol% and increased monotonically with increasing ethanol concentration until 15 vol% at which point the areal expansion reached a plateau of 50%. These observations are consistent with the known physical characteristics of the DPPC interdigitated ($L_{\beta}I$) phase induced by alcohols (as investigated by X-ray diffraction, AFM, and DSC), and hence, our findings are the first direct measurement of the areal expansion that accompanies the $L_{\beta'}$-$L_{\beta}I$ phase transition. For ethanol concentrations between 7 vol% and 15 vol%, areal expansion increased monotonically, consistent with the existence of domains of the $L_\beta I$ and $L_{\beta'}$ phases in coexistence, but for 15 vol% or higher, the observed constant areal expansion indicates that 100% $L_\beta I$ phase exists. At the 7 vol% ethanol threshold concentration, the area expansion exhibited a bimodal distribution of 0% and 20%, indicating that at the threshold concentration, interdigitation is induced in only a portion of DPPC vesicles. Areal expansion could not be easily reversed, consistent with kinetic trapping of the $L_\beta I$ phase. DPPC vesicles exposed to butanol and ethanol exhibited similar areal expansion behaviors, but the threshold concentration for butanol is much lower, 1 vol%, compared to ethanol suggesting that the threshold concentrations (in molar concentration) for short-chain alcohols follows Traube's rule; that is for every additional CH_2 group, three times less alcohol is needed to induce interdigitation. Ethanol-induced areal expansion of DPPC bilayers was inhibited by inclusion of 10 mol% and 25 mol% cholesterol in the bilayer, consistent with improved stability of a liquid-ordered phase. However, areal expansion of DPPC-cholesterol bilayers could be induced by the simultaneous presence of ethanol and application of tension (~8 mN/m), suggestive of a cooperative effect. The presence of 20 vol% ethanol significantly decreased the surface cohesion of DPPC bilayers containing 25 mol% cholesterol as evidenced by a decreased area compressibility modulus and lysis tension. Lastly, these palpable effects on the mechanical properties and area of a lipid membrane in the gel and liquid-ordered phases induced by short-chain alcohols compliment our previous parallel studies on a fluid phase membrane. Given the broad distribution of membrane components, it is highly likely that gel phase, fluid phase, and liquid ordered phase coexist in eukaryotic cell membranes, including those that contain sterols other than cholesterol. These studies demonstrate that significant (and possibly reversible) changes in area per molecule and values of mechanical properties are to be expected in all lipid phases of cell membranes after exposure to concentrations of short-chain alcohols in which major alterations in the metabolism of organisms, especially eukaryotic micro-organisms, have been observed.

Acknowledgements

We would like to thank Professor David Block of the Departments of Chemical Engineering and Materials Science and Viticulture and Enology at the University of California, Davis, for inspiring us in these studies. This work was funded by the Materials

Research Institute at Lawrence Livermore National Laboratory (MI-03-117), the Center for Polymeric Interfaces and Macromolecular Assemblies (Grant NSF DMR 0213618), and the generous endowment from Joe and Essie Smith.

[1] S. A. Simon; T. J. McIntosh, *Biochim. Biophys. Acta* **1984**, *773*, 169-172.
[2] J. L. Slater; C. H. Huang, *Prog. Lipid Res.* **1988**, *27*, 325-359.
[3] L. F. Braganza; D. L. Worcester, *Biochemistry* **1986**, *25*, 2591-2596.
[4] O. P. Bondar; V. G. Pivovarenko; E. S. Rowe, *Biochimica Et Biophysica Acta-Biomembranes* **1998**, *1369*, 119-130.
[5] E. S. Rowe; T. A. Cutrera, *Biochemistry* **1990**, *29*, 10398-10404.
[6] U. Vierl; L. Lobbecke; N. Nagel; G. Cevc, *Biophys. J.* **1994**, *67*, 1067-1079.
[7] N. E. Nagel; G. Cevc; S. Kirchner, *Biochim. Biophys. Acta* **1992**, *1111*, 263-269.
[8] N. A. Avdulov; S. V. Chochina; L. J. Draski; R. A. Deitrich; W. G. Wood, *Alcoholism-Clinical and Experimental Research* **1995**, *19*, 886-891.
[9] J. L. Slater; C. H. Huang, in: "*The Structure of Biological Membranes*"; P., Y., Ed.; CRC Press: Boca Raton, FL, 1992; pp 175-210.
[10] H. Komatsu; S. Okada, *Biochimica Et Biophysica Acta-Biomembranes* **1995**, *1235*, 270-280.
[11] E. T. Kisak; B. Coldren; J. A. Zasadzinski, *Langmuir* **2002**, *18*, 284-288.
[12] H. Komatsu; E. S. Rowe, *Biochemistry* **1991**, *30*, 2463-2470.
[13] P. Laggner; K. Lohner; R. Koynova; B. Tenchov, *Chem. Phys. Lipids* **1991**, *60*, 153-161.
[14] H. V. Ly; M. L. Longo, *In Revision at Biophysical Journal* **2003**.
[15] D. Needham; D. V. Zhelev, in: "*Giant Vesicles*"; Luisi, P. L.; Walde, P., Eds.; John Wiley & Sons, LTD, 2000; Vol. 6, pp 103-147.
[16] D. Needham; D. V. Zhelev, in: "*Vesicles*"; Rosoff, M., Ed.; Marcel Dekker, Inc., 1996; Vol. 62, pp 373-444.
[17] E. A. Evans, *Methods Enzymol.* **1989**, *173*, 3-35.
[18] R. Kwok; E. Evans, *Biophysical Journal* **1981**, *35*, 637-652.
[19] B. M. Discher; Y. Y. Won; D. S. Ege; J. C. M. Lee; F. S. Bates; D. E. Discher; D. A. Hammer, *Science* **1999**, *284*, 1143-1146.
[20] M. L. Longo; A. J. Waring; D. A. Hammer, *Biophysical Journal* **1997**, *73*, 1430-1439.
[21] M. L. Longo; A. J. Waring; L. M. Gordon; D. A. Hammer, *Langmuir* **1998**, *14*, 2385-2395.
[22] D. V. Zhelev; N. Stoicheva; P. Scherrer; D. Needham, *Biophys. J.* **2001**, *81*, 285-304.
[23] D. Needham; D. V. Zhelev, *Annals of Biomedical Engineering* **1995**, *23*, 287-298.
[24] H. V. Ly; D. E. Block; M. L. Longo, *Langmuir* **2002**, *18*, 8988-8995.
[25] J. Silvius, "*Lipid-Protein Interactions*"; John Wiley & Sons: New York, 1982.
[26] D. Needham; E. Evans, *Biochemistry* **1988**, *27*, 8261-8269.
[27] E. Evans; D. Needham, *J. Phys. Chem.* **1987**, *91*, 4219-4228.
[28] W. Rawicz; K. C. Olbrich; T. McIntosh; D. Needham; E. Evans, *Biophys. J.* **2000**, *79*, 328-339.
[29] M. I. Angelova; R. Mutafchieva; R. Dimova; B. Tenchov, *Colloids Surfaces* **1999**, *149*, 201-205.
[30] K. Olbrich; W. Rawicz; D. Needham; E. Evans, *Biophys. J.* **2000**, *79*, 321-327.
[31] B. Deryagin; Y. Gutop, *Kolloidn. Zh.* **1962**, *24*, 370-374.
[32] D. H. Kim; M. J. Costello; P. B. Duncan; D. Needham, *Langmuir* **2003**, *19*, 8455-8466.
[33] E. Evans; F. Ludwig, *Journal of Physics-Condensed Matter* **2000**, *12*, A315-A320.
[34] E. Evans; V. Heinrich; F. Ludwig; W. Rawicz, *Biophys. J.* **2003**, *85*, 2342-2350.
[35] J. C. M. Lee; H. Bermudez; B. M. Discher; M. A. Sheehan; Y.-Y. Won; F. S. Bates; D. E. Discher, *Biotechnol. Bioeng.* **2001**, *73*, 135-145.
[36] E. S. Rowe, *Biochemistry* **1983**, *22*, 3299-3305.
[37] J. Mou; J. Yang; C. Huang; Z. Shao, *Biochemistry* **1994**, *33*, 9981-9985.
[38] S. Maruyama; T. Hata; H. Matsuki; S. Kaneshina, *Biochimica Et Biophysica Acta-Biomembranes* **1997**, *1325*, 272-280.
[39] E. S. Rowe, *Biochim. Biophys. Acta* **1985**, *813*, 321-330.
[40] F. Zhang; E. S. Rowe, *Biochemistry* **1992**, *31*, 2005-2011.
[41] I. Traube, *Ann. Chem. Liebigs* **1891**, *265*, 27-55.
[42] L. Loebbecke; G. Cevc, *Biochim. Biophys. Acta* **1995**, *1237*, 59-69.
[43] E. S. Rowe; J. M. Campion, *Biophys. J.* **1994**, *67*, 1888-1895.
[44] M. F. N. Rosser; H. M. Lu; P. Dea, *Biophys. Chem.* **1999**, *81*, 33-44.

[45] T. P. W. McMullen; R. N. A. H. Lewis; R. N. Mcelhaney, *Biochemistry* **1993**, *32*, 516-522.
[46] S. L. Veatch; S. L. Keller, *Physical Review Letters* **2002**, *8926*, 8101.
[47] J. Z. Lu; Y. H. Hao; J. W. Chen, *J. Biochem. (Tokyo).* **2001**, *129*, 891-898.
[48] E. Endress; S. Bayerl; K. Prechtel; C. Maier; R. Merkel; T. M. Bayerl, *Langmuir* **2002**, *18*, 3293-3299.
[49] D. Needham; R. Nunn, *Biophys. J.* **1990**, *58*, 997-1009.
[50] E. Evans; W. Rawicz, *Physical Review Letters* **1990**, *64*, 2094-2097.
[51] D. Worcester; B. Hammouda, *Physica B* **1997**, *241*, 1175-1177.
[52] D. A. Driscoll; S. Samarasinghe; S. Adamy; J. Jonas; A. Jonas, *Biochemistry* **1991**, *30*, 3322-3327.
[53] H. Ichimori; T. Hata; H. Matsuki; S. Kaneshina, *Biochimica Et Biophysica Acta-Biomembranes* **1998**, *1414*, 165-174.
[54] H. Ichimori; T. Hata; T. Yoshioka; H. Matsuki; S. Kaneshina, *Chem. Phys. Lipids* **1997**, *89*, 97-105.
[55] V. Levadny; V. M. Aguilella; M. Yamazaki, *Chemical Physics Letters* **2002**, *360*, 515-520.
[56] J. W. Zeng; P. L. G. Chong, *Biochemistry* **1991**, *30*, 9485-9491.

Macromol. Symp. **2005**, *219*, 123-133

Spontaneously Forming Unilamellar Phospholipid Vesicles

Baohua Yue,[1] *Chien-Yueh Huang,**[1] *Mu-Ping Nieh,*[2,3] *Charles J. Glinka,*[4] *John Katsaras*[2]

[1]York Department of Chemical Engineering, Department of Chemistry and Environmental Science, New Jersey Institute of Technology, Newark, NJ 07102, USA
E-mail: mhuang@adm.njit.edu
[2]National Research Council, Steacie Institute for Molecular Sciences, Chalk River, Ontario K0J 1J0, Canada
[3]Department of Physics, University of Guelph, Guelph, Ontario N1G-2W1, Canada
[4]Center for Neutron Research, National Institute of Standards and Technology, Gaithersburg, MD 20899, USA

Summary: Unilamellar vesicles (ULV) consisting of a single lipid bilayer are of special interest as drug delivery vehicles. Here, we report on a spontaneously forming ULV system composed of the short- and long-chain phospholipids, dihexanoyl (DHPC) and dimyristoyl (DMPC) phosphorylcholine, respectively, doped with the negatively charged lipid, dimyristoyl phosphorylglycerol (DMPG). Small-angle neutron scattering (SANS) and dynamic light scattering (DLS) were employed to systematically investigate the effects of lipid concentration, salinity, and time on vesicle stability. It is found that ULV size is practically constant over a range of lipid concentration and temperature. The spontaneously formed ULV are stable for periods of four months, or greater, without the use of stabilizers.

Keywords: dynamic light scattering; membranes; neutron scattering; phospholipids; vesicles

Abbreviations: DLS – dynamic light scattering; DHPC – dihexanoyl phosphorylcholine; DMPC – dimyristoyl phosphorylcholine; DMPG – dimyristoyl phosphorylglycerol; MLV – multilamellar vesicle; NG7 – neutron beamline; NIST – National Institute of Standards and Technology; P-B theory – Poisson-Boltzmann theory; PEG – polyethylene glycol; SANS – small angel neutron scattering; ULV – unilamellar vesicle

Introduction

Phospholipid vesicles are considered promising candidates for encapsulation and delivery devices of drugs and biomaterials. Stable, spontaneously forming ULV are of particular interest as there is no need for stabilizers. Nevertheless, spontaneously formed ULV have

 DOI: 10.1002/masy.200550111

predominantly been observed in surfactant systems;[1-13] stable bicompatible phospholipid ULV rarely form spontaneously.

Conventional methods of preparing lipid ULV often involve tedious procedures such as, multi-stage extrusion and sonication, necessary in breaking-up multilamellar vesicles (MLV). However, these ULV are unstable and eventually revert back to MLV.

Some vesicles can be stabilized by polymers, e.g. PEG (polyethylene glycol)-grafting, but the release of the encapsulated materials may be hindered due to low membrane permeability.[14,15] In comparison, spontaneously formed ULV have the possible advantage of high stability, easy preparation, and fast release under appropriate conditions. For drug delivery, of most interest are monodisperse ULV with diameters between 50 and 150 nm, a compromise between loading efficiency, stability and ability to extravasate after administration.[16-17]

Previously, lipid ULV were prepared using long and short-chain lipid mixtures.[18-26] The average vesicle radius $<R_o>$ was found to change with lipid concentration, C_{lp}. A recent study has shown that the structures and sizes of DMPG or Ca^{2+} doped spontaneously formed ULV were practically C_{lp} indepemdent.[27-28] However, when the systems were doped with both DMPG and Ca^{2+}, the ULV radius became C_{lp} dependent. It is known that Ca^{2+} binds tenaciously with the membrane, simultaneously affecting bilayer surface charge and local screening.

In the present study, we replaced Ca^{2+} with Na^{+} in order to alleviate the complexity from surface binding, and to study the effect of ionic strength on spontaneous vesiculation. The self-assembled structures from NaCl-doped DMPC/DMPG/DHPC mixtures were studied using SANS and DLS. The vesicle size was monitored as a function of C_{lp}, solution salinity, C_s, and time t. The surface charge was introduced via the bilayer embedded negatively charged lipid, DMPG.

Under various experimental conditions, it is found that ULV size is insensitive to both C_{lp} and C_s. The formation of vesicles can be explained by the Poisson-Boltzmann theory (P-B theory) of charged fluid membranes, where a negative Gaussian modulus can be obtained[29-31] and spontaneous vesiculation can lead to stable ULV. As salinity increases, the total elastic energy increases and eventually leads to a ULV-MLV transition. The reason for nearly constant vesicle sizes under various lipid concentrations and salinity is presently not very clear. It may result from the dominant role of enthalpic interactions at low lipid concentration or from the kinetics of the system. Despite this, DLS data show most samples remaining stable for weeks and even months.

Experimental Section

Materials: DMPC, DHPC and DMPG were purchased from Avanti Polar Lipids* (Alabaster AL); sodium chloride (NaCl) was obtained from Sigma-Aldrich (St. Louis, MO). All chemicals were used as received. Prior to use, deuterium oxide (99.8 %, Fisher Scientific) solvent was filtered through a 0.1 μm Millipore Millex-VV filter.

Sample Preparation: Solutions of DMPC/DMPG/DHPC molar ratio of 60/1/15 were prepared to a C_{lp} of 1.00 wt% in D_2O (diluting from higher C_{lp} of 5wt.%) by dissolving dry lipid powders. Sodium chloride was then added to yield a C_s in the range of 0.10 % to 1.00 %. After a freezing and thawing cycle for homogenization, samples were diluted with appropriate amounts of D_2O to their final C_{lp} and C_s. Samples were stored at 4 °C for four months, while selected samples were further incubated at 30 °C for an additional month. Most samples were tested with DLS multiple times over a five-month period, while SANS experiments were conducted one month after the sample preparation.

Small Angle Neutron Scattering: SANS experiments were conducted on the 30m NG7 beamline at the NIST Center for Neutron Research, Gaithersburg, MD. Two sample-to-detector distances (1.50 and 15.30 m) were selected, covering an effective Q range between 0.002 and 0.3 A^{-1}.

Dynamic Light Scattering: DLS experiments were performed on a Beckman N4 Plus Photon Correlation Spectrometer equipped with a laser source of wavelength 632.8 nm at a scattering angle of 90°.

Results

All final ULV solutions were transparently with a bluish tinge, a characteristic of solutions containing nanometer sized vesicles.[6] Solutions of MLV, on the other hand, appear opaque and undergo phase separation with time.

Effects of C_{lp} vesicle stability and size: To study the effect of C_{lp} on the self-assembled structures, ULV size for a series of C_{lp} concentrations (e.g., 0.033, 0.10, 0.33, 0.50 and 0.75 wt%, and C_s = 0.1 wt%), were measured.

SANS data and the corresponding best-fit curves of a spherical shell model are shown in

---

The identification of any commercial product or trade name does not imply endorsement or recommendation by the National Institute of Standards and Technology (NIST).

Figure 1. Data of the C_{lp} = 0.75 wt% sample are not fitted because of large particle size and high polydispersity. The fitted results are in good agreement with the experimental data for all samples with $C_{lp} \leq 0.5$ wt%. The results also indicate that all the vesicular structures have a constant bilayer thickness of ~ 32 Å as a function of C_{lp}, consistent with previously reported data.[25-28] Strikingly, vesicle radii remain practically unaltered (24 ± 1 nm) and are C_{lp} independent for $C_{lp} \leq 0.33$ wt%. Larger monodisperse ULV (~ 31 nm) are observed in the C_{lp} = 0.5 wt% sample. The monotonic decay for the 0.75% sample implies the existence of polydisperse large aggregates.

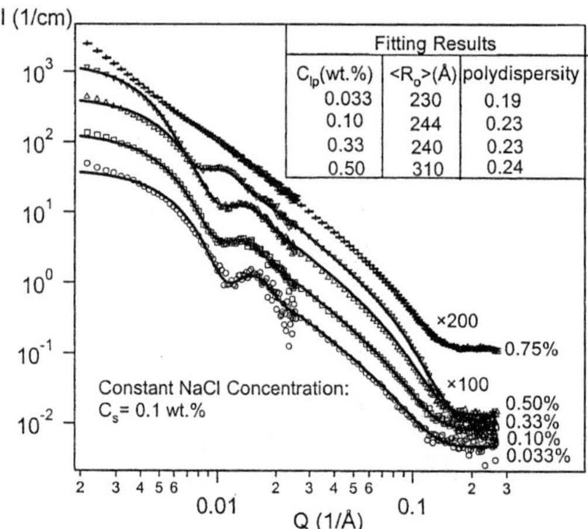

Figure 1. SANS data, fitted curves (solid lines) and results for samples with varying C_{lp} and a constant NaCl concentration of C_s = 0.10 wt%. The C_{lp} = 0.50 and 0.75 wt % samples were scaled for better viewing.

DLS measurements of the same systems (Figure 2) show results consistent with SANS data such as, unimodal size distribution with low polydispersity and a hydrodynamic radius, R_H, ~ 40 nm. As shown in Figure 2, that although particle size remains reasonably constant for three C_{lp} samples (i.e., 0.033, 0.10 and 0.33 wt%), the size distribution for the 0.50 wt% sample is much larger. In the case of the 0.75 wt% sample a bimodal distribution is observed immediately after preparation, with the sample turning turbid. Note that the best-fit vesicle radius from SANS experiments is always smaller than the hydrodynamic radius (R_H) obtained from DLS data. This difference is a result of the different size-averaging algorithms. Moreover, R_H also includes the contribution from associated water molecules, and which is not the case for the radii determined from SANS data. Finally, large aggregates ($R_H > 50$ nm) have been observed in DLS experiments, and may contribute to the misfit of the SANS data at very low Q (< 0.003 $Å^{-1}$) values (Figure 1).

Both SANS and DLS results show that the ULV size is, within a certain C_{lp} range (≤ 0.33 wt%), practically invariant. As C_{lp} increases above 0.5 wt%, the ULV size becomed C_{lp} dependent. Although highly stable ULV were found when doped either with a charged lipid (DMPG) or a salt ($CaCl_2$),[27,28] this is the first time that a highly stable ULV system doped with both charged a lipid (DMPG) and a salt (NaCl), has been reported.

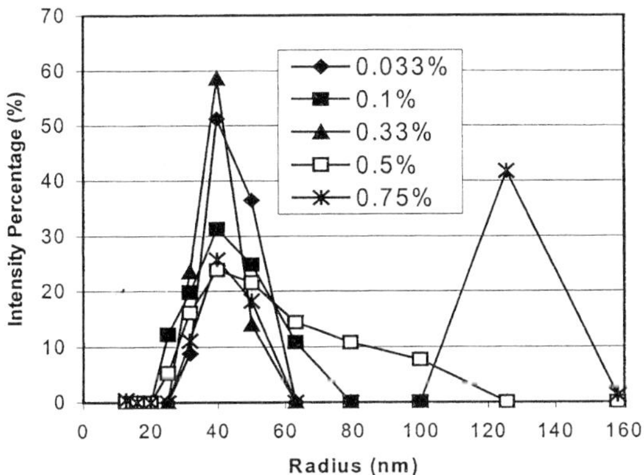

Figure 2. Size distribution functions obtained from DLS data for samples with C_{lp} ranging between 0.033 and 0.75 wt%, at a fixed C_s of 0.10 wt%.

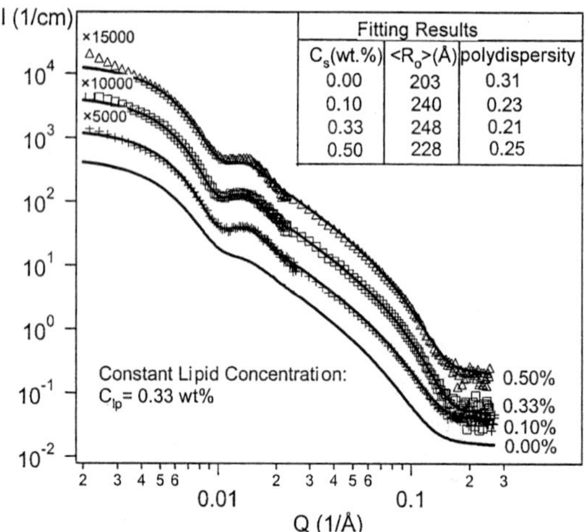

Figure 3. SANS data, fitted curves (solid lines) and results for varying C_s samples and C_{lp} = 0.33 wt%. Data was scaled by an appropriate factor (shown) for better visualization.

Effects of C_s on vesicle stability and size: To study the effect of ionic strength, NaCl was added to vesicle solutions. For a C_{lp} = 0.33 wt% sample, five salt concentrations (C_s = 0, 0.10, 0.33, 0.50 and 1.00 wt%) were examined (Figure 3, except for the C_s = 1.0 wt% sample). Conspicuously, the best fits to the SANS data show the existence of ULV with invariant radii (24 ± 1 nm), and polydispersity ~ 20%, for all samples except those with C_s = 0 and 1 wt%. The sample C_s = 0 wt% sample (non-NaCl doped) has a slightly smaller radius (~ 20 nm) and a slightly higher polydispersity (~ 30%); the C_s = 1.00 wt% sample turned opaque only one day after being prepared, presumably forming MLV.

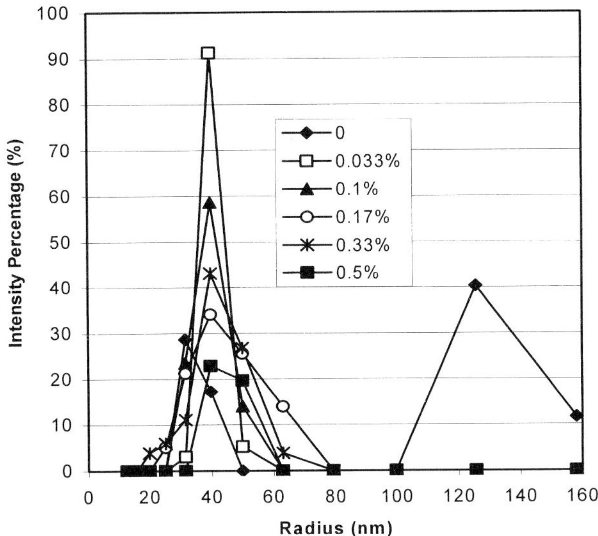

Figure 4. Size distribution functions from DLS for samples with C_s varying from 0.0 to 0.50%, and C_{lp} = 0.33 wt %.

The finding of forming nearly constant vesicle size as a function of C_s is further supported by DLS data (Figure 4), which show ULV radii remainin virtually unaltered at ~ 40 nm. Poydispersity, on the other hand, does seem to be variable. In some cases such as, $C_s = 0$, another peak with a much larger hydrodynamic radius (R_H >100nm) is found. This coexistence of ULV and the large particles may also help in explaining the discrepancy between the SANS data and the fit to the data in very low Q regime (< 0.004 $Å^{-1}$).

Time Evolution: The stability of ULV size versus time was monitored by DLS for four different C_{lp} (0.033, 0.1, 0.33 and 0.5 wt%) samples. All samples were kept at 4 °C for the first four months and at 30 °C during the fifth month. Figure 5 depicts the size changes of the samples: for ULV formed in mixtures with $C_{lp} \leq 0.1$ wt%, R_H were almost invariant over the five month period; for ULV formed with C_{lp}/C_s = 0.33 wt%/0.33 wt%, R_H showed a slight increase in the first four months, followed by a significant increase, as judged from the appearance of a second peak in DLS data, after the fifth month. In the case of C_{lp}/C_s = 0.50 wt%/0.1 wt%, the sample became turbid after two months, indicative of MLV formation.

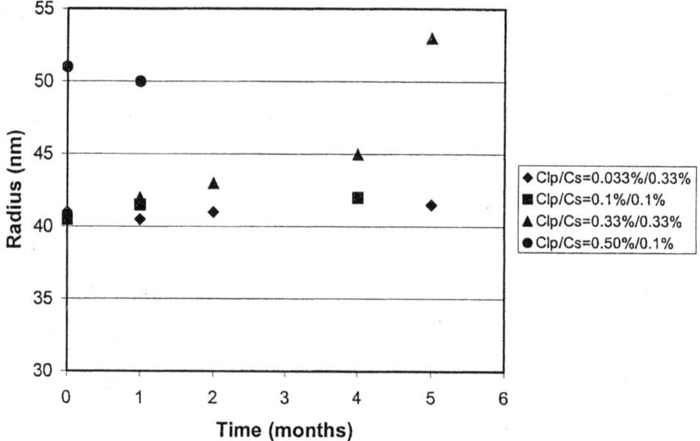

Figure 5. Hydrodynamic radius (R_H) of ULV as a function of time after sample preparation; C_{lp}/C_s = 0.033 wt%/0.33 wt%, 0.10 wt%/0.10 wt%, 0.33 wt%/0.33 wt% and 0.50 wt%/0.10 wt% .

Discussion

Effect of Lipid concentration, C_{lp}: Our observations of variable size ULV for $C_{lp} > 0.33$ wt% samples are consistent with previous studies.[27-28] For $C_{lp} \leq 0.33$ wt% samples, vesicle size is C_{lp}-independent and consistent with what has previously been observed in systems doped with either DMPG or Ca^{2+}, but not with both.[27-28] A previous study of non-doped ([DMPC]/[DMPG] = ∞) and strongly DMPG-doped ([DMPC]/[DMPG] = 15) DMPC/DHPC systems at temperatures and lipid concentrations similar to those studied here, showed the presence of MLV and bicelles, respectively.[26] When an appropriate amount of charged is introduced into DMPC/DHPC mixtures, the mixtures exhibit spontaneous, monodisperse ULV formation . This implies that surface charge may induce a reasonably deep local free energy minimum rendering the ULV insensitive to changes in C_{lp} and C_s. The fixed ratio between charged (DMPG) and non-charged lipids used in this study provides a condition for constant surface charge density assuming that, under the conditions investigated, the amount of lipids in solution does not affect the chemical potential. The insensitivity of ULV size to changes in C_{lp} also contradicts the predictions of Oberdisse et al.,[32-34] whose results show a ULV size increase with increasing surfactant concentration in a system of high surface charge density.

Recently, Egelhaaf et al.[35-36] suggested a kinetic model explaining the formation mechanism of ULV in lecithin/bile salt systems. The model states that ULV could be kinetically trapped as the detergent (i.e., charged short-chain lipid) molecules are removed from the rim of disk-like micelles (i.e., bicelles) upon dilution. Once vesicles are formed, the size may be insensitive to a certain degree of change in either C_s or C_{lp} since only a small amount of bile salt is left residing in lipid membranes. The formation of ULV in our study is different from the above scenario. Although bicelles were found at C_{lp} = 1 wt% and low temperatures[28], the short chain DHPC lipid does not carry any charge and probably distributes itself at the high curvature bicelle rim. This, of course, does not preclude the formation of ULV via another kinetically controlled mechanism. However, more studies on the stability and size of ULV formed *via* various pathways are needed to further clarify this issue.

Effect of Ionic strength: The screening effect of coulombic interactions on the self-assembled charged structures can be studied by altering the ionic strength, e.g., salt concentrations, and thus the Debye length χ_D. Our experimental results show a slightly smaller vesicle radius in NaCl-free mixtures (~ 20 nm, SANS) compared to those ULV formed from lipid mixtures with NaCl (~ 24 nm, SANS). Assuming the complete dissociation of DMPG in a NaCl-free mixture, the calculated χ_D is about 37 nm, larger than the radius of curvature, and the surface charge density of the bilayer membrane is about 3.5×10^{-3} C/m^2 (neglecting the charge difference between the inner and outer layers and taking vesicle outer radius R_o = 24 nm and inner radius R_i = 21 nm). Compared to the calculations from the simplified P-B theory by Winterhalter and Helfrich, [31] the above experimentally obtained charge density and the Debye length predict a small positive bending modulus and a negative Gaussian modulus, leading to a negative total elastic energy which favors spontaneous vesiculation. Reducing the screen lengths (by increasing salt concentration) renders a reduction in bending modulus, however, under the given charge density, the negative Gaussian modulus *increases* with the ionic strength and eventually leads to a positive total energy at certain threshold $C_{s,MLV}$, where a ULV to MLV transition takes place. For samples doped with NaCl, χ_D varies from 4 nm (C_s = 0.033 wt%, 5.6 mM) to 1.3 nm (C_s = 0.33 wt%, 56 mM), ULV size remains unaltered. Though there is no direct comparison between our data and calculations from more detailed treatments such as, P-B theory with translational entropy, asymmetric charge density, etc., the insensitivity of ULV size may be attributed to a low entropic contribution resulting from low lipid concentration and the dominant role of enthalpic interactions in forming self-assembled structures. As lipid concentration is increased, other degrees of freedom should be considered, particularly their contributions to entropy. The entropic contribution should increase with the total lipid concentration that eventually destabilizes ULV at a threshold salinity $C_{s,MLV}$, which decreases with the total lipid concentration.

Aging Effect: Although the size of ULV, over a five month period, is found to be invariant (Figure 5), the issue of whether these ULV are kinetically trapped or thermodynamically stable is not yet understood. Note that the samples were stored at 4 ^{o}C, lower than the T_M of DMPC, except during the measurements. Most of the time, the DMPC was in the gel phase, a more rigid molecular structure than the L_α phase, and hence the exchange rate of DMPC from ULV to solution, or vice versa, was slow. The invariant ULV size at low temperature (4 ^{o}C) may suggest that the vesicular structures are kinetically trapped.

Conclusions

Spontaneously formed ULV with low polydispersity are found in DMPC/DMPG/DHPC phospholipids mixtures by dilution from high concentration lipid solution at low temperature. SANS and DLS results show that the ULV are reasonably stable over prolonged periods of time and variations of lipid concentration (C_{lp} < 0.33%) and ionic strength (C_s < 0.5 wt%), both below and above the phase transition temperature, T_M, of DMPC. These results are different from some theoretical predictions and experimental reports. A possible explanation is that the enthalpic contribution from a negative Gaussian modulus dominates over the effects from other degrees of freedom at low lipid concentrations. Another possible explanation may be the energetically favorable packing of DMPC and DMPG that renders a slow response to any change in chemical potential at low concentrations. Further studies of the resultant structures *via* various pathways in preparation with the same chemical composition are needed to better understand the formation mechanism of ULV. The stability of ULV and the ability to control their size may facilitate the development of pharmaceutical or biomedical applications.

Acknowledgment

The authors would like to thank the instrument scientists at NCNR NG7 for their assistance. BY thanks the travel support of the outreach program by the University of

Maryland for conducting the neutron experiment. MPN and JK appreciate the many fruitful discussions with Dr. V. A. Rughunathan (Raman Research Institute, India).

[1] Kaler, E. W.; Murthy, A. K.; Rodrigues, B. E.; Zasadzinski, J. A. *Science* **1989**, *245*, 1371.
[2] Kaler, E. W.; Herrington, K. L.; Murthy, A. K.; Zasadzinski, J. A. *J. Phys. Chem.* **1992**, *96*, 6698.
[3] Murthy, A. K.; Kaler, E. W.; Zasadzinski, J. A. *J. Colloid and Interface Sci.* **1991**, *145*, 598.
[4] Hoffmann, H.; Thunig, C.; Munkert, U. *Langmuir* **1992**, *8*, 2629
[5] Hoffmann, H.; Munkert, U.; Thunig, C; Valinente, M. *J. Colloid and Interface Sci.* **1994**, *163(1)*, 217.
[6] Yatcilla, M. T.; Herrington, K. L.; Brasher, L. L.; Kaler, E. W.; Chiruvolu, S.; Zasadzinski, J. A. *J. Phys. Chem.* **1996**, *100*, 5874.
[7] Bergstrom, M. *Langmuir* **1996**, *12*, 2454.
[8] Bergstrom, M.; Eriksson, J. C. *Langmuir* **1996**, *12*, 624.
[9] Bergstrom, M.; Eriksson, J. C. *Langmuir* **1998**, *14*, 288.
[10] Bergstrom, M.; Pedersen, J. S.; Schurtenberg, P.; Egelhaaf, S. U. *J. Phys. Chem.B* **1999**, *103*, 9888.
[11] Bergstrom, M.; Pedersen, J. S. *J. Phys. Chem. B* **2000**, *104*, 4155.
[12] Bergstrom, M.; *J. Colloid Interface Sci.* **2001**, *240*, 294.
[13] Viseu, M. I.; Edwards, K.; Campos, C. S.; Costa, S. M. *Langmuir* **2000**, *16*, 2105.
[14] Papahadjopoulos, D.; Vail, W. J.; Newton, C.; Nir, S.; Jacobson, K.; Poste, G.; Lazo, R. Biochim. Biophys. Acta **1977**, *465*, 579.
[15] Barenholz, Y. *Curr. Opin. Coll.* **2001**, *6*, 66.
[16] Lasic, D. *TIBTECH* **1998**, *16*, 307.
[17] Lasic, D. D.; Joannic, R.; Keller, B. C.; Frederik, P. M.; Auvray, L. Adv. Colloid Sci. **2001**, *89-90*, 337.
[18] Gabriel, N. E.; Roberts, M. F. *Biochemistry.* **1984**, *23,* 4011.
[19] Ollivon, M.; Lesieur; S.; Grabrielle-Madelmont, C.; Paternotre, M. *Biochim. Biophys. Acta.* **2000**, *1508,* 34.
[20] Schurtenberger, P.; Mazer, N.; Waldvogel, S.; Kanzig, W. *Biochim. Biophys. Acta.* **1984**, *775*, 111.
[21] Schurtenberger, P.; Mazer, N.; Kanzig, W. *J. Phys. Chem.* **1985**, *89*, 1042.
[22] Egelhaaf, S. U.; Schurtenberger, P. *Phys. Rev. Lett.* **1999**, *82*, 2804.
[23] Andelman, D.; Kozlov, M. M.; Helfrich, W. Europhys. Lett. 1994, 25, 231.
[24] Lesieur, P.; Kiselev, M. A.; Barsukov, L. I.; Lombardo, D. *J. Appl. Cryst.* **2000**, *33*, 623.
[25] Nieh, M.-P.; Glinka, C. J.; Krueger, S.; Prosser, R. S.; Katsaras, J. *Langmuir* **2001**, *17*, 2629.
[26] Nieh, M.-P.; Glinka, C. J.; Krueger, S.; Prosser, R. S.; Katsaras, J. *Biophys. J.* **2002**, *82*, 2487.
[27] Nieh, M.-P.; Harroun, T. A.; Raghunathan, V. A.; Glinka, C. J.; Katsaras, J. *Phys. Rev. Lett.* **2003**, *91*, 158105.
[28] Nieh, M.-P.; Harroun, T. A.; Raghunathan, V. A.; Glinka, C. J.; Katsaras, J. *Biophys. J.* **2004**, *86*, 2615.
[29] Helfrich, W. Z. *Naturforsch* **1973**, *28c*, 693-703.
[30] Winterhalter, M.; Helfrich W. *J. Phys. Chem.* **1988**, 92, 8685.
[31] Winterhalter, M.; Helfrich W. *J. Phys. Chem.* **1992**, 96, 327.
[32] Oberdisse, J.; Couve, C.; Appell, J.; Berret, J. F.; Ligoure, C.; Porte, G. *Langmuir* **1996**, *12*, 1212.
[33] Oberdisse, J.; Porte, G. *Phys. Rev. E.* **1997**, *56*, 1965.
[34] Oberdisse, J. *Euro. Phys. J. B.* **1998**, *3*, 463.
[35] Leng, J.; Egelhaaf, U.; Cates, M. E. *Europhys. Lett.* **2002**, *59 (2)*, 311.
[36] Leng, J.; Egelhaaf, U.; Cates, M. E. *Biophys. J.* **2003**, *85*, 1624.

Structural Phase Behavior of High-Concentration, Alignable Biomimetic Bicelle Mixtures

Mu-Ping Nieh,[*1,2] V. A. Raghunathan,[3] Charles J. Glinka,[4] Thad Harroun,[1] John Katsaras[1]*

[1]National Research Council, Steacie Institute for Molecular Sciences, Neutron Program for Materials Research, Chalk River, Ontario K0J 1J0, Canada
E-mail: mu-ping.nieh@nrc.gc.ca
[2]Department of Physics, University of Guelph, Guelph, Ontario N1G-2W1, Canada
[3]Raman Research Institute, Bangalore, 560 080, India
[4]Center for Neutron Research, National Institute of Standards and Technology, Gaithersburg, MD 20899, USA

Summary: The structures of bicelle mixtures composed of dimyristoyl and dihexanoyl phosphatidylcholines (DMPC and DHPC) with DMPC/DHPC molar ratios of 3.2 and 5 are characterized using polarized optical microscopy (POM) and small angle neutron scattering (SANS). Three phases, isotropic (I), chiral nematic (N^*) and smectic (S) are observed as temperature (T) varies from 10 to 70 °C. The structure of the magnetically alignable N^* phase, which was previously considered to be made up of discoidal micelles, is found to be composed of “ribbons”. Doping with the charged lipid, dimyristoyl phosphatidylglycerol (DMPG), which has the same 14:0 hydrocarbon chains as DMPC, results in a structural change of the aggregates where only the isotropic and smectic phases are observed. The smectic phase for the mixtures doped with DMPG is shear-alignable and follows one-dimensional swelling. However, at high-T zwitterionic DMPC/DHPC mixtures form multi-lamellar vesicles (MLV) with a relatively constant lamellar spacing of 66 Å, independent of water content.

Keywords: bicelle; DHPC; DMPC; DMPG; lamellae; multilamellar vesicle (MLV); neutron diffraction; optical microscopy; phospholipids; ribbons; small angle neutron scattering (SANS)

Introduction

Bicelle mixtures have been extensively used as magnetically alignable membrane substrates for the study of membrane-associated proteins/peptides.[1-3] Their chemical compositions include at least one long-chain and one short-chain lipid/surfactant molecule.[4-6] From nuclear magnetic resonance (NMR) studies a discoidal structure has been inferred with the short-chain lipid sequestered at the rim of the bilayered micelle, or so-called “bicelle”, and the long-chain lipid forming the planar bilayer. Because the edge energy at the disk’s rim can be reduced by the above-mentioned lipid arrangement, the proposed structure has been widely accepted. Recently, another possible structure for the

 DOI: 10.1002/masy.200550112

magnetically alignable phase, namely perforated lamellae, was proposed based on SANS data. From the data, it was inferred that the short-chain lipids coat the rims of the perforations, minimizing the pore curvature energy.[7,8] In order to understand the structures and their self-assembly mechanisms, we have investigated a series of high-concentration bicelle mixtures at two different long- to short-chain lipid molar ratios, Q (=3.2 and 5), using POM and SANS. At least three phases (isotropic, chiral nematic and smectic) were observed with increasing temperature (e.g., 10 to 70 °C).

In general, the lipid bilayer normal (N) tends to align perpendicular to the applied magnetic field (B) due to the negative diamagnetic susceptibility of the lipids. When doped with certain lanthanide ions (Tm^{3+}, Yb^{3+}, etc.),[9-11] the bilayers can align with N || B, eliminating the degree of freedom for the rotation of N. These lanthanide ion-doped bicelles mixtures are popular because they can serve as a "biological goniometer" for membrane-associated proteins/peptides, making it possible to obtain their in-plane or out-of-plane structure. However, the non-biologically relevant lanthanide ions may bind with the protein/peptide of interest, altering its native conformation. Moreover, a strong magnetic field is required to align the bicelle mixture resulting, with the exception of NMR, in a nontrivial experimental setup.

Here we report a bicelle formulation that is alignable under a weak shear flow in absence of lanthanide ions and strong magnetic fields. The "alignability" is comparable to those systems oriented in the presence of strong magnetic fields.[12]

Experimental Section

DMPC, DHPC and dimyristoyl phosphatidylglycerol (DMPG) were purchased from Avanti Polar Lipids[†] and used without further purification. The mixtures were dissolved in D_2O (99.9% in purity, Chalk River Lab.) at a total lipid concentration = 25 wt.% and two Q values, 3.2 and 5. DMPG-doped mixtures with Q = 5 were prepared at different molar ratios of DMPG to DMPC, [DMPG]/[DMPC] = 0.02 and 1, yielding different charge densities.

The POM studies were performed on an Olympus BX52 microscope and the images were captured with a CCD camera. Samples loaded in sealed rectangular capillaries (2mm X 0.1mm, VitroCom Inc.), were placed in a temperature-controlled stage (Linkam

---

[†] The identification of any commercial product or trade name does not imply endorsement or recommendation by the National Institute of Standards and Technology

TMHS600) to obtain the desired T with an accuracy of ± 0.05 °C. Each sample was equilibrated at the desired T for at least 0.5 hours before the image was recorded.

SANS experiments were performed at the 30-m SANS instrument (NG7) located at the National Institute of Standards and Technology (NIST, Gaithersburg, Maryland, USA). Two sample-to-detector distances (1.25 and 15.3 m) were selected to cover a range of scattering vectors, q, (= $4\pi/\lambda \sin\frac{\theta}{2}$, where λ and θ are the wavelength of the neutron and the scattering angle, respectively) from 0.002 to 0.35 $Å^{-1}$. Samples were loaded in cells with a path length of 2 mm and placed in a temperature controlled sample holder capable of handling 10 samples. Temperature was controlled by a circulating water bath. Each measurement was taken after the sample was equilibrated at the desired T for 0.5 hours. Two-dimentional SANS data were corrected for background (blocked beam), and reduced to an absolute intensity scale based on the incident neutron flux. The data were finally circularly averaged to yield one-dimentsional SANS data.[13]

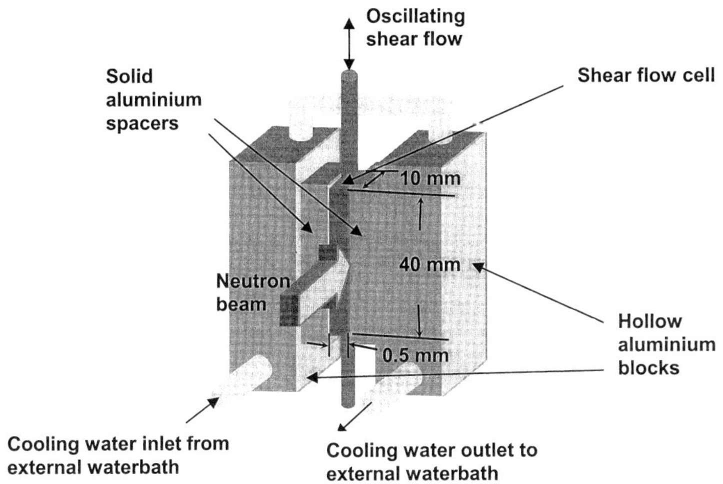

Figure 1. Neutron diffraction experimental setup for shear-aligned samples.

Neutron diffraction experiments were conducted at the N5 diffractometer located at the National Research Universal (NRU) reactor (Chalk River Laboratories, Ontario, Canada). 2.37 Å neutrons were selected using the (002) reflection of a pyrolytic–graphite monochromator, while a graphite filter was used for the scattered beam to reduce diffraction arising from $\lambda/2$ and $\lambda/3$ incident neutrons. Samples were aligned using an oscillating shear flow along the long axis of a rectangular cell with openings at both ends (0.5mm X 10mm X 40mm, Starna Cells, Inc.). After shear, the normal of the alignable bilayers remains perpendicular to the 10mm X 40mm surfaces for extended periods of time (at least several days). Sample cells were sandwiched between two aluminium blocks, whose temperature was controlled using a circulating waterbath (Figure 1). The incident beam was parallel to the 10mm dimension of the cell. To determine the degree of sample alignment, rocking curves were obtained using the first-order Bragg peak position.

Results and Discussion

Non-doped DMPC/DHPC mixtures. Q = 3.2 and 5 bicelle mixtures have a similar temperature dependence for appearance and viscosity. At low T (< 20 °C), the samples are more or less transparent with water-like viscosity (Figure 2A). When T is increased slightly above 23 °C (phase transition temperature, T_M, of DMPC), the sample's appearance is either transparent (Q = 3.2) or translucent (Q=5), and both become highly viscous (Figure 2B). On further elevation of T, the samples turn opaque and experience a drop in viscosity (Figure 2C). Our observation is consistent with those in the literature, where the viscosity of the systems is reported to have a maximum at T ~ 30 °C.[14,15] The POM results of the Q = 3.2 mixture identify three distinct textures, namely smectic, chiral nematic and isotropic phases on decreasing T. In the high-T smectic phase, the bright four-fold brushes of a "Maltese cross" are often observed (Figure 3A). This texture is characteristic of multi-lamellar vesicles coexisting with the isotropic phase, a phase that has no birefringence (dark regions). With decreasing T, a S → N^* phase transition is observed where POM data show a "fingerprint" texture (Figure 3B).[16] Moreover, the high viscosity N^* phase (Figure 2B) indicates the presence of entangled aggregates, difficult to justify with a discoidal structure. The S → N^* transition temperature, $T_{S\to N^*}$, is also strongly Q-dependent ($T_{S\to N^*}$ = 30 and 52 °C for Q = 5 and 3.2, respectively), indicating the importance of DHPC. Decreasing T below the T_M of DMPC induces a N^* → I transition, where POM shows no birefringence intensity. Unlike the $T_{S\to N^*}$, the N^* → I transition temperature, $T_{N^*\to I}$, is independent of Q and coincides with the T_M of DMPC (23 °C).

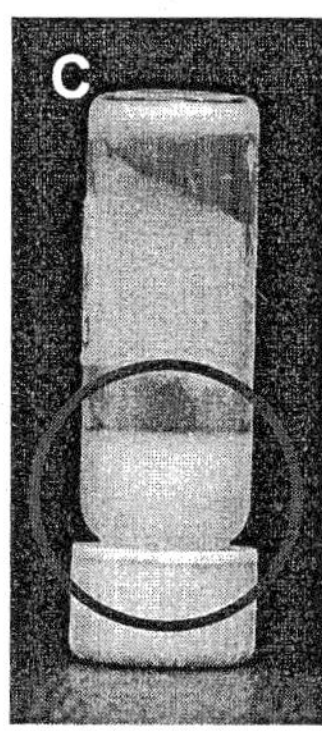

Figure 2. Appearance and viscosity of the 25 wt.% Q = 5 sample at (A) T = 10 °C, where the low viscosity sample is transparent, (B) at T = 25 °C , where the highly viscous sample is translucent and finally, (C) at 50 °C, where the sample experiences a drop in viscosity and turns opaque. Blue circles indicate the sample's location with respect to the glass vial.

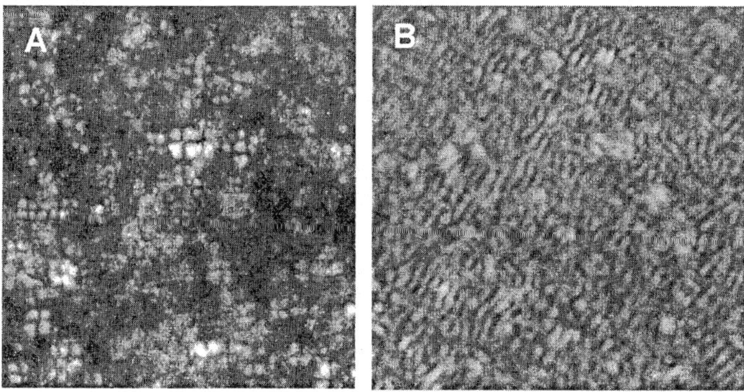

Figure 3. POM textures of a 25 wt.% Q = 3.2 sample (A) at T = 70 °C showing a "Maltese cross" texture, and representative of MLV. (B) "Fingerprint" texture at T = 45 °C characteristic of an N^* phase.

The microscopic structures of these three phases were characterized using SANS. Figure 4 shows SANS data of a Q = 3.2 sample at 3 different temperatures. It should be noted that the Q = 5 sample exhibits a similar behaviour as the Q = 3.2 sample (data not shown). At high T (70 °C), the sharp peak at q = 0.095 $Å^{-1}$ is indicative of a smectic phase with a lamellar spacing of ~ 66 Å, consistent with fully hydrated DMPC bilayers and the POM data showing MLV (Figure 3A). In the low-q regime, the intensity follows a q^{-4} decay, characteristic of large MLV of the order of hundreds of nm (Porod law). This result is also consistent with the opaqueness seen in these samples, and DHPC phase separating from from DMPC, as reported in several NMR studies.[17,18]

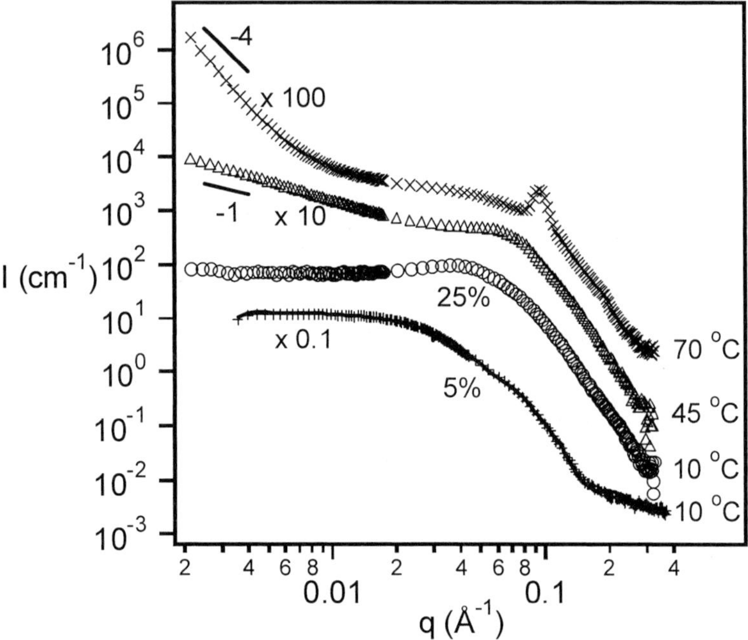

Figure 4. SANS results for the 25 wt.% Q = 3.2 sample at three different temperatures (70, 45 and 10 °C represented by x, Δ and o, respectively). SANS data for a 5 wt.% dilute solution (+) and its best-fit using a discoidal model (solid curve), are also shown. Structure factor was not taken into account when fitting the data.

At T = 45 °C, a less well-defined peak is found at q = 0.05 - 0.06 $Å^{-1}$, indicative of a poorly ordered structure. At this temperature the POM data shows the presence of N^* phase (Figure 3B). This phase was previously found to be magnetically alignable with the bilayer normal ⊥ to the applied magnetic field, and the proposed morphology was purported to be a discoidal bilayered micelle, or commonly referred-to bicelle.[2,4-6] The low-q SANS intensity shows a q^{-1} decay over a decade of q (from 0.002 to 0.02 $Å^{-1}$), representative of long objects (e.g., cylinders or ribbons). This holds true assuming that the contribution from the structure factor does not alter the slope over this q range.[16] Judging from numerous NMR experiments carried-out in this phase, the consensus is that some of the DHPC phase separates from DMPC.[1-6] Combining the SANS and NMR evidence along with the high viscosity of the solution, it therefore seems likely that the morphology is one of entangled ribbon-like objects with DMPC found on the flat surface of the ribbons and DHPC coating the ribbon's rim, thus minimizing the structure's edge energy. From the SANS data the average minimal persistence length and the ribbon width can be estimated to be > 3000 and 300Å, respectively. Most recently, cryo-transmitted electron microscopy of a similar system also shows enlongated aggregates over this T range.[19]

As mentioned previously, $T_{S \rightarrow N^*}$ is strongly Q-dependent: the larger the Q (i.e., lower short-chain composition) the narrower the range of the N^* phase. Decreasing T, presumably results in more DHPC molecules at the edge, either due to lower solubility of DHPC in water, or to the two lipids phase separating. The extra DHPC molecules break up the MLV structure while stabilizing the ribbons by coating the rim of the ribbons and minimizing the edge energy. Therefore, the $T_{S \rightarrow N^*}$ is lower for higher Q value samples (i.e., less DHPC). The reason for this is that a sufficient amount of DHPC needs to be released either from bilayers or solution in order stabilize the ribbons.

At T = 10 °C, the SANS result shows a weak interparticle interaction peak at q ~ 0.04 $Å^{-1}$ and a plateau region at low q values (Figure 4). SANS data from a lower lipid concentration (5 wt.%) sample, where no such peak is observed, can be best-fitted using a disk model confirming that the isotropic phase consists of discoidal aggregates, similar to bicelles. The best-fit diameter and thickness of the resultant bicelles are ~ 160 and 50 Å, respectively. Since $T_{N^* \rightarrow I}$ always occurs around the T_M of DMPC for both Q = 3.2 and Q = 5 samples, the implication is that the formation of bicelles requires an ordered DMPC gel phase. It is expected that more DHPC is needed to coat the rim of a bicelle than the rim of a ribbon because the bilayer area of an individual ribbon is much larger than that of an individual disk. A possible interpretation is that L_α-phase DHPC is highly immiscible with gel-phase DMPC, resulting in almost all of the DHPC being driven out from the bilayer region. As a result, ribbons break up into the smaller bicelles in order to accommodate the extra DHPC. However, more experiments are needed to further elucidate this proposal.

In summary, three structures (bicelles, ribbons and MLV) are found in DMPC/DHPC zwitterionic mixtures as a function of T. The location and number density of DHPC molecules in the structure presumably play an important role in determining the structural phase of the mixtures.

DMPG-doped DMPC/DHPC mixtures. Neutron diffraction experiments[12] and NMR spectroscopic studies[10] have shown that doping the zwitterionic DMPC/DHPC mixtures with paramagnetic ions (e.g., Tm^{3+}) results in a nematic → smectic phase transition. In our experiments, instead of using lanthanide ions, we doped the mixtures with a small amount of biologically relevant DMPG to yield a membrane with good alignment properties after the application of shear. A previous study has shown that doping DMPC/DHPC mixtures with DMPG results in the alignment of lamellar domains in two distinct orientations.[20] Presently, this phenomenon is not well-understood, and the study of charged systems may provide us with better insight of the influence of charge density on the morphology of bicelle mixtures.

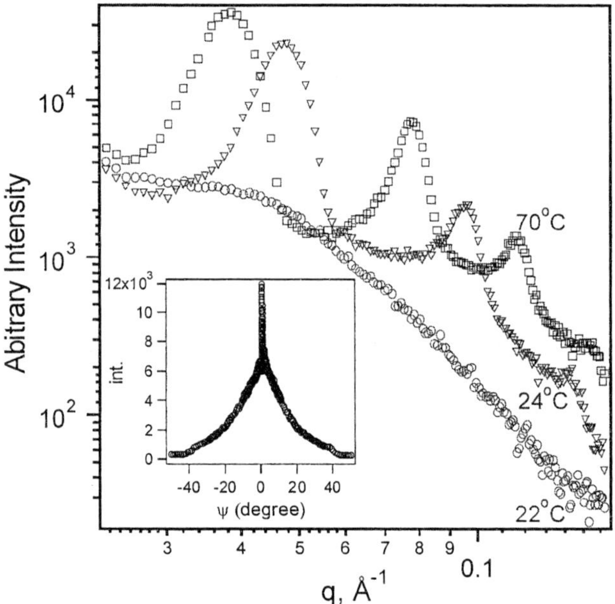

Figure 5. Neutron diffraction patterns of a 20 wt.% Q = 5 and [DMPG]/[DMPC] = 0.02 sample at three temperatures (22, 24 and 70 °C represented by o, Δ and □, respectively). A bicelle → lamellar transition is observed at T ~ 23 °C. The inset is the "rocking curve" at the first order Bragg position, q_{max} for T = 70 °C.

Figure 5 shows the neutron diffraction data for a 20 wt.% Q = 5 and [DMPG]/[DMPC] = 0.02 sample using the experimental setup shown in Figure 1. The data indicate a bicelle (isotropic) → lamellar (smectic) transition taking place at ~ 23 °C (~ T_M of DMPC). This lamellar phase persists until 70 °C (Figure 5). From previous studies we know that diluting the system to 5 wt.% results in one-dimensional swelling, characteristic of a lamellar phase without excess water.[8] This behaviour is very different from that of zwitterionic MLV, whose lamellar repeat spacing is insensitive to dilution. Apparently, even this small amount of DMPG is more than enough to result in "complete unbinding", as recently reported by Demé et al. on dioleoylphosphatidylserine solutions.[21,22] The fact that the spacing of these lamellae increases with T from 133 Å (24 °C) to 165 Å (70 °C), is presumably the result of perforations, present in the low T phase, annealing at high T.[23]

The degree of sample alignment can be obtained from the "rocking curve", which shows the scattered intensity at specific q_o (usually at a peak position, q_{max}) as a function of sample angle, ψ, with respect to the incident neutron beam. The intensity corresponds to the population of lamellae with a spacing $2\pi/q_o$ oriented at $(\psi - \phi_o/2)$ with the confining surface, where ϕ_o is the scattering angle of the first order Bragg reflection. The inset to Figure 5 shows a spike at $\psi_o = \phi_o/2$ with a full width at half maximum (FWHM) of less than 1° and broad wings spanning over ± 40°. The spike represents a portion of lamellae that align with the confining surfaces, while the broad wings are the result of a population of misaligned lamellae which deviate from ψ_o. The alignment of DMPG-doped lamellae is found to be dependant on the membrane charge density in a manner that the peak intensity of the rocking curve for a sample with a higher membrane charge density is higher, indicative of more lamellae aligning with the confining surfaces.

We have also found that the alignability of DMPG-doped samples ([DMPG]/[DMPC] = 1) after a weak shear is comparable with, if not better than, that of Tm^{3+}-doped membranes aligned in a strong magnetic field (2.6 Tesla),[12] as shown in Figure 6. The DMPG-doped mixture has a larger lamellar spacing than the previously studied Tm^{3+}-doped system (~ 23 wt.%),[12] due to a slightly lower lipid content. Up to six Bragg reflections from the shear aligned sample are observed.

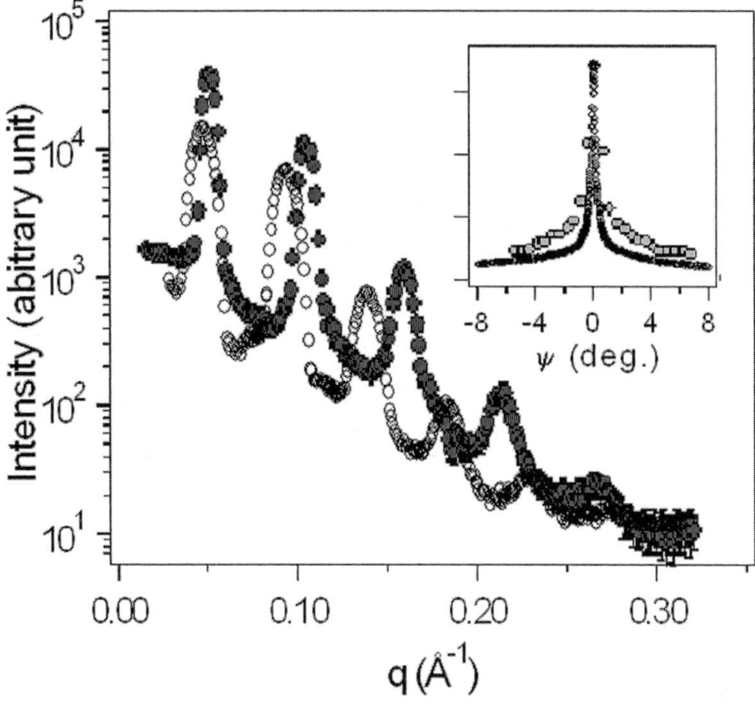

Figure 6. Neutron diffraction patterns of Tm^{3+}-doped mixtures (23 wt.%) in magnetic field ~ 2.6 Tesla (filled symbols)[12] and DMPG-doped mixtures (20 wt.%, [DMPG]/[DMPG] = 1) after experiencing a weak shear (open symbols). The inset to the figure shows the rocking curves of the two aligned samples.

Conclusion

We have shown that DMPC/DHPC mixtures (Q = 3.2 and 5) undergo isotropic (bicelle) → nematic (ribbon) → smectic (MLV) transitions with increasing T. $T_{I \rightarrow N*}$ is tied to the T_M of DMPC, whereas $T_{N* \rightarrow S}$ is strongly Q-dependant. The magnetically alignable nematic phase seems to be made up of ribbons, instead of bicelles as thought previously. Bicelles, however, form the low-T isotropic phase, while MLV are the resultant structure at high T. For DMPG-doped systems no N^* (ribbon) phase is found; only the I → S transition is observed with $T_{I \rightarrow S}$ taking place around the T_M of DMPC. Also, unlike excess water zwitterionic MLV, the smectic phase for DMPG-doped mixtures can swell up to several hundred Å. DMPG-doped systems can yield a comparable degree of alignment, after the application of a weak oscillating shear, as the previously studied Tm^{3+}-doped systems in a strong magnetic field. The DMPG-doped mixtures can potentially be used as a "biological goniometer" to study the in-plane and out-of-plane structure of proteins and peptides associated with biomembranes without the use of lanthanide ions and strong magnetic fields.

Acknowledgement

This work utilized facilities supported in part by the National Science Foundation under Agreement No. DMR-9986442. J. K. and M-P. N. thank the Advanced Food and Materials Network (Network of Centres of Excellence, Canada) and the Neutron Program for Materials Research, (National Research Council) for financial support.

[1] C. R. Sanders II, G. C. Landis, *J. Am. Chem. Soc.* **1994**, *116*, 6470.
[2] C. R. Sanders II, G. C. Landis, *Biochemistry* **1995**, *34*, 4030.
[3] R. S. Prosser, V. B. Volkov, I. V. Shiyanovskaya *Biophy. J.* **1998**, *75*, 2163.
[4] C. R. Sanders II, J. H. Prestegard, *Biophys. J.* **1990**, 58, 447.
[5] C. R. Sanders II, J. P. Schwonek, *Biochemistry.* **1992**, *31,* 8898.
[6] C. R. Sanders II, B. J. Hare, K. P. Howard, J. H. Prestegard, *Prog. NMR Spectro.* **1994**, *26*, 421.
[7] M.-P. Nieh, C. J. Glinka, S. Krueger, R. S. Prosser, J. Katsaras, *Langmuir* **2001.** *17*, 2629.
[8] M.-P. Nieh, C. J. Glinka, S. Krueger, R. S. Prosser, J. Katsaras, *Biophys. J.* **2002.** *82*, 2487.
[9] R. S. Prosser, S. A. Hunt, J. A. DiNatale, R. R. Vold, *J. Am. Chem. Soc.* **1996.** *118*, 269.
[10] R. S. Prosser, J. S. Hwang, R. R. Vold, *Biophys. J.* **1998.** *74*, 2405.
[11] R. S. Prosser, V. B. Volkov, I. V. Shiyanovskaya *Biochem. Cell Biol.* **1998**, *76*, 443.
[12] J. Katsaras, R. L. Donaberger, I. P. Swainson, D. C.Tennant, Z. Tun, R. R. Vold, R. S. Prosser, *Phys. Rev. Lett.* **1997**, *78*, 899.
[13] C. J Glinka, J. G. Barker, B. Hammouda, S. Krueger, J. J. Moyer, W. J. Orts. *J. Appl. Cryst.* **1998.** *31*, 430.
[14] J. Struppe, R. R. Vold, *J. Magn. Reson.* **1998.** *135*, 541.
[15] J. S. Hwang, G. A. Oweimreen, *Arab. J. Sci. Eng.* **2003.** *28*, 43.
[16] M.-P. Nieh, V. A. Raghunathan, C. J. Glinka, T. A. Thad, G. Pabst, J. Katsaras, *Langmuir* (*in press).*
[17] G. Raffard, S. Steinbruckner, A. Arnold, J. H. Davis, E. J. Dufourc, *Langmuir* **2000.** *16*, 7655.
[18] E. Sternin, D. Nizza, K Gawrisch, *Langmuir* **2001.** *17*, 2610.
[19] L. van Dam, G. Karlsson, K. Edwards, *Biochim. Biophys. Acta* **2004.** *1664*, 241.
[20] M.-P. Nieh, V. A. Raghunathan, H. Wang, J. Katsaras, *Langmuir* **2003.** *19*, 6936.
[21] B. Demé, M. Dubois, T. Gulik-Krzywicki, T. Zemb, *Langmuir* **2003.** *18*, 997.
[22] B. Demé, M. Dubois, T. Zemb, *Langmuir* **2003.** *18*, 1005.
[23] M.-P. Nieh, V. A. Raghunathan, G. Pabst, T. A. Thad, J. Katsaras, *submitted to J. Phys. Chem.*

Calcium-Alginate Nanoparticles Formed by Reverse Microemulsion as Gene Carriers

Jin-Oh You,[1] *Ching-An Peng**[1,2]

[1]Department of Chemical Engineering, University of Southern California, Los Angeles, CA 90089, USA
[2]Department of Materials Science, University of Southern California, Los Angeles, CA 90089, USA
E-mail: capeng@usc.edu

Summary: Natural biopolymers are widely used in the field of drug and gene delivery. In this study, alginate nanoparticles were prepared using water-in-oil microemulsion as a template followed by calcium crosslinking of guluronic acid units of alginate polymer. After collected by ultracentrifugation, alginate nanoparticles were analyzed by electron microscopy to obtain the size and morphology which were varied with the ratio of water, oil, and surfactant used. To examine the potency of Ca-alginate nanoparticles as carriers for gene delivery, GFP-encoding plasmids were encapsulated in these nanoparticles to investigate the degree of endocytosis by NIH 3T3 cells and ensuing transfection rate. Our results showed that Ca-alginate nanoparticles with an average size around 80 nm in diameter are very efficient gene carriers, in comparison with plasmid DNA condensed by polyethyleneimine (PEI).

Keywords: AOT; calcium-alginate nanoparticle; gene delivery; polyethyleneimine; reverse microemulsion

Introduction

It has been shown that the potential risks (e.g., immunogenicity and oncogenicity) of using viral vectors for gene therapy are obstacles for clinical use even though virus-mediated gene delivery offers high transduction efficiency. As a result, much attention has been focused on the design of synthetic cationic vectors for nucleic acid condensation and delivery. Despite the promise of developing clinically useful nonviral vectors, most nonviral gene carriers provide less efficient gene transfer, especially in escape from endosomal vesicles. Although quite a few endosomolytic chemicals (e.g., polyethyleneimine [1], chloroquine [2], and cationic lipid [3]) have been used to facilitate endosomal escape of plasmid DNA, cytotoxicity induced by these agents is a major obstacle need to tackle with. In this study, we propose to use Ca-alginate nanoparticles as the gene delivery vehicle since it is biocompatible and biodegradable. Moreover, once internalized via endocytosis pathways, they can undergo quick erosion and elicit osmotic swelling, hence facilitating endosomal escape of gene to the cytoplasm.

 DOI: 10.1002/masy.200550113

In order to make alginate-based nanoparticles, we employed self-assembly microemulsion as the template. The water-in-oil (w/o) microemulsion system is thermodynamically stable; typically the composition is 10, 5, and 85% by volume of the water, surfactant, and oil, respectively. Water-in-oil microemulsions generally contain nanometer-sized water droplets stabilized by a curved surfactant monolayer. To date, a good range of different microemulsion systems has been used, and polymerization of the continuous, dispersed and surfactant phase has been attempted.[4,5] Daubresse et al. have described an interesting approach for immobilizing enzymes in nanoparticles prepared from microemulsions.[6] Both nonbiodegradable poly(acrylamide) and biodegradable poly(acryldextran) nanoparticles were prepared using w/o microemulsions stabilized by a mixture of Aerosol-OT (AOT; sodium bis(2-ethylhexyl) sulfosuccinate) and Brij 30. Addition of the enzyme alkaline phosphatase to the reaction mixture caused the enzyme to become entrapped in the nanoparticles. The resulting particles were typically between 25 and 50 nm, with a polydispersity of ~ 20%. However, to the best of our knowledge, no study has yet demonstrated the capability of making nanometer-scale Ca-alginate particles using water-in-oil microemulsion as the template.

For the reverse microemulsion developed here, toluene was used as the oil phase, AOT was the surfactant, and 0.5% sodium alginate (Na-alginate) solution was the aqueous phase. AOT is one of the most investigated anionic surfactants. Because of its double-tailed nature, the AOT molecule is almost structurally balanced with respect to its hydrophilic and hydrophobic parts and is one of the few ionic surfactants that can form microemulsions without the addition of a co-surfactant.[7-10] After washing AOT from the reverse microemulsion template by toluene, the structure of Ca-alginate nanoparticles has a core of Na-alginate and an enclosed membrane formed by gelation of Na-alginate using calcium chloride. For the encapsulation of plasmid DNA, plasmids encoded with enhanced green fluorescent protein (eGFP) were pre-mixed with sodium alginate solution and then added dropwise into AOT/toluene mixture. The formed DNA/alginate nanocontainers were gelled by the addition of calcium chloride. The formed DNA/alginate nanoparticles were cultured with NIH 3T3 cells and led to 50% cells with the expression of eGFP post 48-h incubation, which is comparable to the result done by PEI/DNA complexes.

Materials and Methods

Preparation of Ca-alginate Nanoparticles

First, 0.25 g of AOT (sodium bis(2-ethylhexyl) sulfosuccinate, Sigma) was dissolved into a vial pre-filled with 13.11 g of toluene. Then, 0.5% sodium alginate solution (medium viscosity; viscosity of 2% solution is approximately 3,500 cps at 25°C, Sigma) was added dropwise into the vial with a micropipette. Various phase appearance (from transparent to turbid) was obtained during the process of adding alginate solution. For each distinct phase, the mixture was vortexed and waited for 10 minutes prior to the addition of 2 mL of filtered calcium chloride solution (2% by weight) for obtaining alginate-based nanodroplets by crosslinking. After carefully washed with acetone and deionized water to remove all residual AOT and toluene, the vial was centrifuged (33,000 ×g) for 30 min and a small white pellet of material was obtained. Finally, the prepared nanoparticles were re-suspended with deionized water. The procedure of encapsulating plasmid DNA into Ca-alginate nanoparticles was the same as the one just described above, except pDNA was pre-mixed with 0.5% sodium alginate solution.

Characterization of Ca-alginate Nanoparticles

The size and morphology of prepared Ca-alginate nanoparticles were analyzed by transmission electron microscopy (model 420, Philips, Eindhoven, Netherlands) and scanning electron microscopy (model S-570, Hitachi, Tokyo, Japan). For TEM images, aqueous solution of the nanoparticles was dropped onto a copper grid (200 mesh) supporting a thin film of amorphous carbon. The excess liquid was swept away with a filter paper, and the grid was dried in air. For SEM images, a drop of aqueous solution containing Ca-alginate nanoparticles was placed on a specimen stub and coated with gold-palladium by sputtering for 2 minutes with 50 mA (K-550X, Emitech LTD, Kent, England).

Transfection of NIH 3T3 Cells

NIH 3T3 cells in the exponential growth phase were detached with 1X EDTA-trypsin (Irvine Scientific, Santa Ana, CA) from a culture dish (Falcon, Franklin Lakes, NJ). 5×10^4 of NIH 3T3 cells were inoculated on a plastic petri dish (35 mm in diameter) with 2 mL of high-glucose Dulbecco's modified Eagle medium (DMEM, Mediatech Inc., VA) supplemented with 10% fetal bovine serum (FBS, Irvine Scientific). After 24-h incubation, Ca-alginate nanoparticles encapsulated with 2 μg plasmid DNA encoding eGFP (pIRES2-EGFP vector, Clontech, Palo Alto, CA) were suspended in 2 mL of growth medium and layered on top of the NIH 3T3 cells which were attached on the bottom of the culture dish. NIH 3T3 cells exposed to pDNA encapsulated Ca-alginate nanoparticles were incubated for transfection. All cell cultures were performed at 37°C and balanced with 5% CO_2 in a 100% humidified incubator. For comparison, transfection of NIH 3T3 cells was also performed using PEI/DNA complexes (MW of PEI = 750 kDa; N/P ratio = 6, where N is the amount of amino nitrogens of PEI and P stands for the amount of phosphate groups of the plasmid DNA).

Results and Discussion

Determination of Reverse Microemulsion Domain

In order to obtain particles with a diameter close or less than 100 nm, water soluble sodium alginate was introduced into diluted water-in-oil microemulsion stabilized by AOT surfactant. Since sodium alginate is insoluble in the oil (toluene), the alginate polymer is confined within the aqueous nanophase. The phase stability of the resulting microemulsion is significantly affected by the amount of added alginate solution. As shown in Figure 1, point A (toluene:alginate solution:AOT = 86.1:0.6:13.3 by weight) in the phase diagram domain indicates the region of transparent microemulsion. Upon addition of sufficient 0.5% alginate solution, the mixture was transformed into translucent (point B) then milk-like appearance (point C).

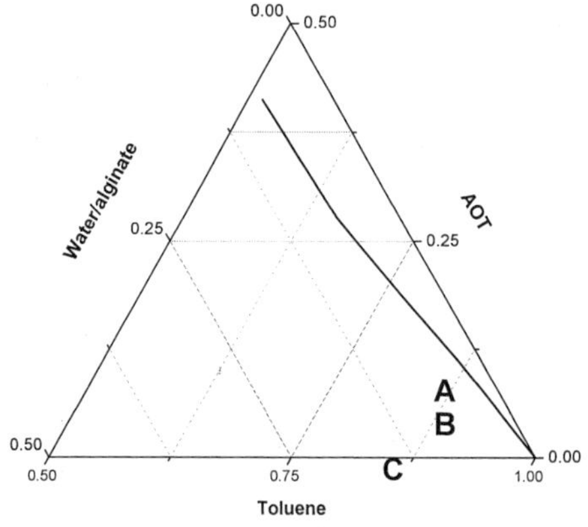

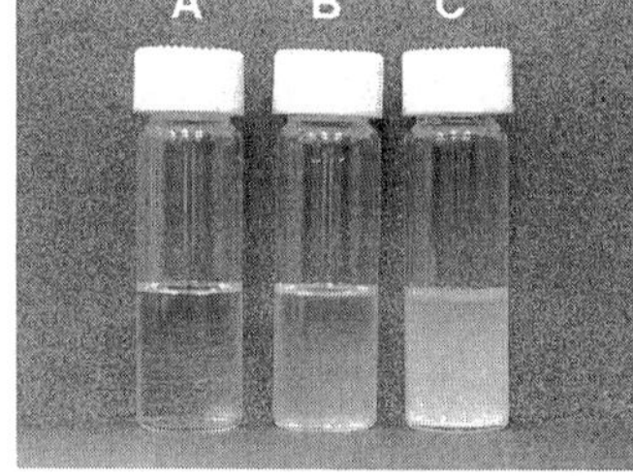

Figure 1. Phase diagram for toluene/alginate solution/AOT (micro)emulsion. Points A (toluene/alginate solution/AOT = 86.1:0.6:13.3 by weight) and C (toluene/alginate solution/AOT = 81:10.8:8.2 by weight) refer to compositions used for the preparation of alginate microemulsion and emulsion. Point A corresponds to the w/o microemulsion region, while, point C is in the emulsion region. Point B (toluene/alginate solution/AOT = 85.7:1.8:12.5 by weight) is located on the boundary line distinguishing the domains of microemulsion and emulsion.

Size and Morphology of Ca-alginate Nanoparticles

The alginate microemulsion at point A in the phase diagram was further mixed with calcium chloride solution to form Ca-alginate nanoparticles. Figure 2 shows the TEM and SEM images of Ca-alginate nanoparticles obtained from the template of microemulsion, boundary, and emulsion, respectively. Figure 2 (A) illustrates separated and spherical nanoparticles with the size ranged from 55 nm to 100 nm. In Figure 2 (B), necklace-like Ca-alginate nanoparticles were identified. More specifically, the size of each single nanoparticle was bigger than that of nanoparticles formed by the microemulsion template. We speculate that alginate biopolymer may interfere with the interdroplet correlations, particularly if it is too large to fit into a single droplet and may thus require a cluster of several droplets for accommodation. To be exact, the biopolymer is covered with a linear array of discrete droplets so as to make up a necklace. In Figure 2 (C), prepared Ca-alginate particles further aggregate with rough morphology. For polymer chains too long to be accommodated in a single particle, necklaces may be formed as shown above. In order to prevent potential aggregation of alginate nanoparticles in the process of gelation within w/o microemulsion template, alginate polysaccharides probably need to be trimmed down to small molecular weights.

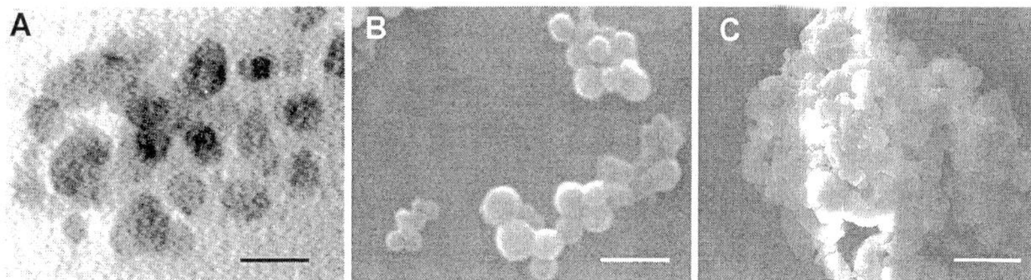

Figure 2. Images of Ca-alginate particles. a) TEM image nanoparticles formed from point A in the microemulsion zone of the phase diagram given in Figure 1 (scale bar = 100 nm; b) SEM image of nanoparticles obtained from point B in the boundary zone (scale bar = 300 nm); c) SEM image of particle aggregates by calcium crosslinking of point C in the emulsion zone (scale bar = 1 μm).

Transfection Rate Enhanced by Ca-Alginate Nanoparticles Encapsulated with pDNA

To examine the efficacy of Ca-alginate nanoparticles in facilitating endosomal escape of pDNA, NIH 3T3 cells were incubated with nanoparticles obtained from the microemulsion template (i.e., point A of Figure 1) for 24, 48 and 72 h, respectively. Percentages of eGFP-expressing NIH 3T3 cells were determined by both bright-field and fluorescent microscopy (see Figure 3-I). For comparison, transfection of NIH 3T3 cells was also performed using PEI/DNA complex (see Figure 3-II) which has been known for its capability of enhancing pDNA release from endosomes. As shown in Figure 3, the discrepancy of transfection rate between using PEI/DNA complexes (53%) and pDNA encapsulation in Ca-alginate nanoparticles (8.5%) was prominent after 24-h incubation. However, after 48-h incubation, the transfection rate using Ca-alginate nanoparticles encapsulated with pDNA was greatly increased up to 48%, in comparison with the one using PEI/DNA complexes (55%). The reason causing such dramatic enhancement of transfection rate is probably because endosomal swelling, elicited by Ca-alginate nanoparticles, reaches its peak after 48-h incubation, hence expediting the release of encapsulated pDNA into cytosol. Transfection rates were also quantitatively determined by measuring the intensity of eGFP expression in cells using a fluorimetric microplate reader with an excitation wavelength of 472 nm and emission wavelength of 512 nm (shown in Figure 4). The trends of this result were consistent with the one given in Figure 3.

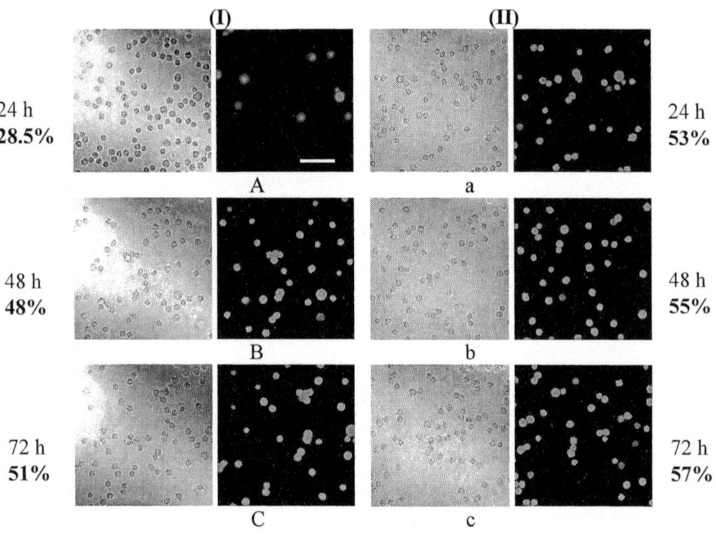

Figure 3. Photomicrographical images of eGFP-expressing NIH 3T3 cells under both bright-field and fluorescent microscopy after been incubated with (I) pDNA encapsulated in Ca-alginate nanoparticles and (II) pDNA condensed with PEI for 24 h (A, a), 48 h (B, b), and 72 h (C, c), respectively. These images represent one set of triplicate experimental data. Note the cells were suspended by trypsinization before the images were taken. Scale bar = 50 μm.

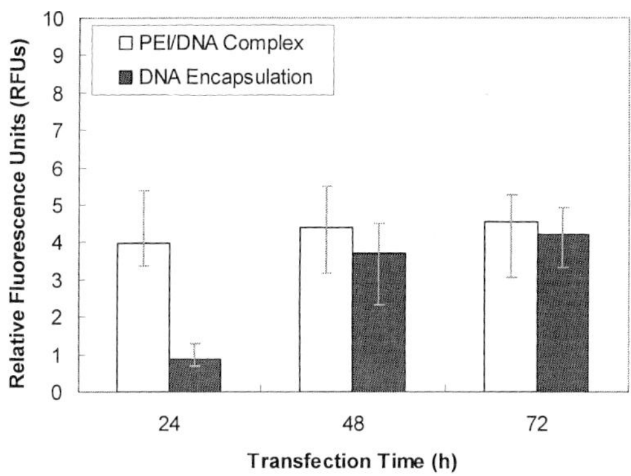

Figure 4. Transfection efficiency of NIH 3T3 cells treated separately with PEI/DNA complex and Ca-alginate nanoparticles encapsulated with pDNA for various periods of time. The relative fluorescence units of eGFP-expressing NIH 3T3 cells were detected by a fluorimetric microplate reader. All the experimental data were obtained in triplicate.

Conclusion

The size of Ca-alginate nanoparticles was significantly modulated by the ratio of toluene, AOT, and sodium alginate solution. Ca-alginate nanoparticles with an average size of 80 nm were able to encapsulate plasmid DNA and ferry the gene into non-phagocytic cells via endocytosis pathway. It seems that the alginate-based gene carriers can assist DNA escape within 1-2 day timeframe from the endosomal lumen into the cytoplasmic space, leading to high transfection efficiency.

[1] O. Boussif, F. Lezoualch, M. A. Zanta, M. D. Mergny, D. Scherman, B. Demeneix, J. Behr, *Proc. Natl. Acad. Sci. USA* **1995**, 92, 7297.

[2] J. D. Fritz, H. Herweijer, G. Zhang, J. A. Wolff, *Hum. Gene Ther.* **1996**, 7, 1395.

[3] D. D. Lasic, N. S. Templeton, *Adv. Drug Deliv. Rev.* **1996**, 20, 221.

[4] J.O. Stoffer, T. Bone, *J. Polym. Sci. Polym. Chem. Ed.* **1980**, 18, 2641.

[5] L.M. Gan, T.D. Li, C.H. Chew, W.K. Teo, L.H. Gan, *Langmuir* **1995**, 11, 3316.

[6] C. Daubresse, C. Grandfils, R. Jerome, P. Teyssie, *J. Colloid Interface Sci.* **1994**, 168, 222.

[7] J. T. Brooks, M. E. Cates, *Chem. Phys.* **1990**, 149, 97.

[8] W. Sager, *Langmuir.* **1998**, 14, 6385.

[9] E. Acosta, D. H. Kurlat, M. Bisceglia, B. Ginzberg, L. Baikauskas, S. D. Romano, *Colloids and surfaces A;Physicochemical and Engineering Aspects.* **1996**, 106, 11.

[10] T. Banerjee, S. Mitra, A. K. Singh, R. K. Sharma, A. Maitra, *Int. J. Pharm.* **2002**, 243, 93.